Fundamentals of Optical Fibers

Fundamentals of Optical Fibers

JOHN A. BUCK
School of Electrical and Computer Engineering
Georgia Institute of Technology

A Wiley-Interscience Publication

John Wiley & Sons, Inc.

New York / Chichester / Brisbane / Toronto / Singapore

Copyright © 1995 by John Wiley & Sons, Inc.

Library of Congress Cataloging in Publication Data:
Buck, John A.
    Fundamentals of optical fibers / John A. Buck.
      p. cm. — (Wiley series in pure and applied optics)
    "A Wiley-Interscience Publication."
    Includes index.
    ISBN 0-471-30818-8 (acid-free paper)
    1. Fiber optics. 2. Optical fibers. 3. Optical communications.
    4. Optical wave guides. I. Title. II. Series.
    TA1800.B83   1995
    621.36′92—dc20   94-27481

To the Memory of My Father
Norton B. Buck
1923–1988

Contents

Preface

This book is an intermediate-level field theory text on optical fiber waveguides. It is intended for use by students at the advanced senior to first-year graduate levels in electrical engineering or in physics who have had at least a one-semester course in electromagnetics at the junior or senior levels. As the title indicates, its purpose is to provide the reader with an account of the *basics* of light propagation in fibers; these include fundamental waveguiding concepts, the influence of various fiber structures and material properties on light transmission, and, finally, nonlinear effects.

Since the primary application of optical fibers is to communication systems, the topics were selected on the basis of their relevance to communications problems. An "undercurrent" of issues and perspectives that pertain to fiber systems is present throughout the treatment. Consequently, in addition to its primary function as a text for courses on fiber fundamentals or on optical waveguiding, it is hoped that the book may serve in a supplementary role to the many existing fiber systems books that are in current use in communications-oriented courses.

The selection of material to be covered was influenced by two considerations: (1) the material should be relevant to present and future needs, while (2) each topic should serve as a vehicle for increased physical understanding. Although the main emphasis is on fibers, such a study has the added benefit of exposing the student to many important concepts in optics that are applicable elsewhere. These considerations have defined the overall structure of the book, as well as the course in which its manuscript has been used in various developmental stages over the past six years.

The book is optimally suited to a one-semester course. For a one-quarter (ten-week) course, Chapters 1 through 6 could be covered, or Chapter 6 could be omitted and material in Chapter 8 could be presented, using Chapter 7 as a reference. The problems at the end of each chapter are in most cases designed to carry on the discussion by exposing the student to special cases or additional interpretations.

I am indebted to several people who have provided many useful comments and suggestions on the manuscript for this book. These include Andrew Dienes (University of California at Davis), William B. Gardner (AT&T Bell Laboratories), R. Steven Weis (Texas Christian University), John R. Whinnery (University of California at Berkeley) and many anonymous reviewers who have provided feedback on various parts of the manuscript throughout its development. In addition, I thank the students who have taken my course over the years, too numerous to name, who have come to me with suggestions or questions that have resulted in improved approaches to ways of explaining the material. Finally, I acknowledge the invaluable help of Michael Tidwell, Ivan Lee, and Darryl Cross in creating the numerous graphs and field plots.

JOHN A. BUCK

Atlanta, Georgia
August 1994

Introduction

The importance of optical fibers in today's world is impossible to overemphasize. What began as a few basic experiments in signal transmission in the 1960s has emerged as a multibillion dollar industry, establishing fiber-based communication networks around the world. As of this writing, fiber systems having information capacities of 2.5 Gbits/s (gigabits per second) are in place, with 10 Gbits/s systems soon to be available. Fiber systems find use anywhere from small isolated networks within a single building or facility, to gigantic intercontinental systems, that include transoceanic links.

There are several advantages for using optical fibers as transmission media. Many of these result from the fact that the principal material in fiber manufacture is glass. Glass, in addition to being abundant and inexpensive, enables low-loss propagation at optical frequencies; the use of optical carrier frequencies (in the vicinity of 10^{14} Hz) establishes the intrinsic possibility of achieving bandwidths on that order. Other considerations that will be discussed in fact limit bandwidths in fibers to values that are substantially less than this, but the frequency range over which good fibers exhibit low losses is in fact on the order of 10^{13} Hz. A further advantage of glass is that it is non-conductive; the fiber thus forms a transmission channel that is essentially immune to most forms of external electromagnetic interference.

Another important advantage of the fiber waveguide is that it presents a small cross-sectional area in the plane transverse to the propagation direction. The fact that this area is much smaller than that of conventional transmission lines has greatly contributed to the economic feasibility of replacing coaxial line-based systems with fiber-based systems, for the simple reason that existing underground conduits (having limited cross-sections) can be used. Consequently the motivation for making the change is high, in that within an existing support structure, changing to fiber transmission significantly increases the information capacity for a given conduit size, while realizing the other advantages of low noise, and the possibility of lower losses.

The primary method of transmitting information via optical fiber is digital—that is, by sequences of optical pulses whose positions, widths, or occurrences can be modulated. The simplest case is binary "on–off" pulse code modulation, in which the presence or absence of pulses within equally spaced time slots (one pulse per slot) corresponds to the occurrences of ones or zeros in the message, respectively. The information capacity of the fiber link can be measured by how many time slots (or bits) can occur per second. This number is limited by (1) the maximum switching speed of the modulator at the input end, (2) the speed of the detector and receiver at the output end, and (3) loss and dispersive effects in the fiber, which degrade the signal and increase the probability of error in detection. An example of this is the mechanism of *group velocity dispersion* in single-mode fibers, which has the effect of broadening the pulses as they propagate. The broadened pulses are reduced in peak power and spread into adjacent time slots, thus increasing the uncertainty that the receiver will correctly interpret the energy within a given time slot as a one or a zero. Smaller, more closely spaced time slots require shorter pulses, which in turn (as will be explained) exhibit greater dispersive broadening. In fact, it is dispersion that provides the most serious bandwidth limitation in fiber systems. A major effort in the development of optical fibers has been in establishing designs that simultaneously minimize loss and dispersion.

Dispersive pulse broadening and loss both increase in proportion to the length of the link. Consequently, repeater stations are positioned at appropriate intervals over long links, for the purpose of detecting, electrically amplifying, filtering, and then regenerating the optical signal. The dispersion and loss experienced over the previous span are thus effectively reset to zero before retransmission. Repeaters are by nature complicated, and so their presence increases the likelihood of system problems arising from component failure. Although individual repeater units are not very expensive, they are expensive to house, and must often be located in areas that are difficult to access. Consequently there is much incentive to avoid using repeaters altogether, or at least to minimize their number by using large spacing.

Optical amplification has recently been introduced as an alternative to the electronic repeater, providing increased simplicity and greater flexibility on choice of location and carrier wavelength. A special case is the fiber amplifier, which is constructed by doping conventional fiber with rare-earth ions, such as erbium. The resulting structure will provide gain for light at the signal wavelength in the presence of strong additional light at a shorter wavelength; the latter "pump" light is provided at the amplifier location. The effect of using a fiber amplifier as a repeater is that losses experienced by the signal are compensated, but the pulse broadening effects of dispersion are not. It is possible to compensate dispersion through one of many passive equalization schemes or devices. The increased simplicity arises from the fact that the signal is maintained in optical form throughout the link, rather than being continually converted between optical and electronic form at each repeater. Additionally, in a wavelength division multiplexed system, it is possible for a single fiber amplifier to boost the power in all wavelengths simultaneously, provided all occur within the amplifier emission spectrum. In contrast, in systems employing electronic repeaters, a separate repeater is needed for each wavelength.

The expansion of fiber networks to encompass larger areas, in addition to increases in repeater spacing, means that higher optical power levels are needed. Increasing power results in the occurrence of various nonlinear processes in the fibers, which can have either positive or negative effects on the system performance. Examples of these effects include *stimulated Raman scattering*, which leads to the generation of light of a longer wavelength than the signal, and which is a mechanism by which cross-talk can occur between signals at two different wavelengths. Another effect is *self-phase modulation*, which results in the broadening of the frequency spectrum of a pulse, while producing a frequency sweep over the pulse width. This process can be used to effectively compensate dispersion, yielding a pulse, known as an *optical soliton*, which propagates through the fiber without changing its shape. Various other nonlinear effects will occur, an understanding of which is essential in the design of modern systems.

Other uses of fiber include optical circuitry, for the processing and storage of data, and sensor applications. The latter include the detection of dimensional changes that could occur in buildings or in aircraft, for example, which incorporate fibers within their structures. Fiber probes are used as sensors of environmental conditions such as temperature and pressure. Most of these schemes are based on the measurement of shifts of interference fringe patterns that occur as a result of minute changes in the length of a fiber; the fiber serves as one arm of an optical interferometer.

The text can be divided roughly into three parts. The first, consisting of Chapters 1 through 3, develops the field equations and basic principles of dielectric guides. Beginning with a review of the pertinent electromagnetic principles in Chapter 1, Chapter 2 then applies these concepts to analyze wave propagation in the symmetric slab waveguide. This treatment serves the purpose of introducing the basic behavior of light in dielectric guides that will be qualitatively similar to that found in fibers. Chapter 3 treats the step index optical fiber using the techniques that were introduced in Chapter 2. The chapter concludes with coverage of single-mode fiber parameters and measurement methods.

The second part of the text, Chapters 4 through 6, deals with limitations in fiber transmission—most notably loss and dispersion—and methods that are used to minimize these problems. Since many loss and dispersion issues are closely tied to fabrication procedures, these are treated briefly in Chapter 4. Basic dispersion concepts and their fiber manifestations are developed in Chapter 5, particularly as they relate to the propagation of pulses and the limitations on signal bandwidth. Chapter 6 explores the use of alternative fiber designs as motivated by the issues presented in Chapters 4 and 5. Field analysis methods for single-mode and multimode fibers, presented in Chapter 6, build on the techniques that were presented in Chapter 3.

The third part of the text, consisting of Chapters 7 and 8, discusses the origins of nonlinear effects and the conditions under which they appear in fibers. Chapter 7 provides a background in nonlinear optics that concentrates on the primary effects that occur in fibers. The treatment is classical and relies on simple oscillator models to develop the nonlinear susceptibilities. Chapter 8 is designed to stand alone in presenting the fiber nonlinear effects, while referring to Chapter 7 for derivations of the equations that are presented. A brief treatment of fiber amplifiers and a presentation of the basic principles of light amplification by stimulated emission are also given in Chapter 8.

1

Selected Topics in Electromagnetic Wave Propagation

The study of optical fiber waveguides, and all other electromagnetic phenomena, begins with Maxwell's equations. These provide the mathematical foundation that is used to model and describe the flow of energy in all types of guiding structures. This chapter is specifically concerned with the fundamental concepts and properties of time-harmonic electromagnetic waves. Such waves consist of electric and magnetic fields that are solutions of Maxwell's equations, subject to the specific material properties and boundary conditions that define the media through which they propagate. The emphasis here is on principles that are to be used in the chapters that follow. The references at the end of the chapter are recommended for additional review as necessary.

We begin with the statement and description of Maxwell's equations. These are expressed below in point form, using MKS units:

$$\nabla \times \mathbf{E} = -\frac{\partial \mathbf{B}}{\partial t} \tag{1.1}$$

$$\nabla \times \mathbf{H} = \mathbf{J} + \frac{\partial \mathbf{D}}{\partial t} \tag{1.2}$$

$$\nabla \cdot \mathbf{D} = \rho_v \tag{1.3}$$

$$\nabla \cdot \mathbf{B} = 0 \tag{1.4}$$

The fields appearing in the equations include the electric field strength, $\mathbf{E}$ (volts/meter, V/m), and the electric displacement, $\mathbf{D}$ (coulombs/meter^2, C/m^2). The two fields are related by way of the material properties through:

$$\mathbf{D} = \epsilon \mathbf{E} = \epsilon_0 \mathbf{E} + \mathbf{P} \tag{1.5}$$

where ϵ is the permittivity of the medium, expressed in units of farads/meter (F/m); ϵ_0 is the free-space permittivity, equal to 8.85×10^{-12} F/m. $\mathbf{P}$ is the *polarization density field* for the medium, interpreted as the dipole moment per unit volume, expressed in units of C/m^2. The polarization, induced by the applied electric field, appears as an additional field that must be accounted for in determining the net electric field, $\mathbf{E}$. The relation between $\mathbf{E}$ and $\mathbf{P}$ is through the *electric susceptiblility function*, χ_e:

$$\mathbf{P} = \epsilon_0 \chi_e \mathbf{E} \tag{1.6}$$

Substitution of (1.6) into (1.5) yields the simplified relation, $\mathbf{D} = \epsilon_0 \epsilon_r \mathbf{E}$, where the dielectric constant ϵ_r is defined as $1 + \chi_e$. χ_e, and thus ϵ_r, are in general tensors, in which case the medium they describe is *anisotropic*. At optical frequencies the medium refractive index, $n = \sqrt{\epsilon_r}$, is usually used to specify the dielectric constant. The net permittivity of the medium, ϵ, is written as the product: $\epsilon = \epsilon_r \epsilon_0 = n^2 \epsilon_0$.

In the present treatment (and in the subsequent four chapters), waves are assumed to propagate in *simple media*, which are by definition *linear, homogeneous,* and *isotropic*. In linear media, χ_e is independent of field strength, a condition that will be lifted in Chapters 7 and 8. In homogeneous media, χ_e is independent of position within the material. Finally, in isotropic materials, χ_e is independent of field orientation and is thus a scalar, as opposed to a tensor.

Relations similar to those above apply to the other field quantities that appear in (1.1) through (1.4). Specifically, the magnetic induction, $\mathbf{B}$ (webers/meter^2, W/m^2), is related to the magnetic field strength, $\mathbf{H}$ (amperes/meter, A/m) through $\mathbf{B} = \mu \mathbf{H}$, where μ (henries/meter, H/m) is the permeability tensor. In free space, the (scalar) permeability is $\mu_0 = 4\pi \times 10^{-7}$ H/m. The net permeability of a material can be expressed in terms of the relative permeability, μ_r, and the free-space value through $\mu = \mu_r \mu_0$. Finally, volume current density, $\mathbf{J}$ (A/m^2), is related to electric field strength through the conductivity, σ: $\mathbf{J} = \sigma \mathbf{E}$.

1.1. ELECTROMAGNETIC WAVE EQUATIONS IN SOURCELESS MEDIA

The materials and structures under consideration throughout this book are assumed sourceless, so that conductivity, σ, and free charge, expressed through the volume charge density, ρ_v, are both zero. The wave equation for the electric field is derived by first taking the curl of (1.1). Equation (1.2) with $\mathbf{J} = 0$ is then substituted into the result to yield:

$$\nabla \times \nabla \times \mathbf{E} = -\mu \frac{\partial}{\partial t} \nabla \times \mathbf{H} = -\mu\epsilon \frac{\partial^2 \mathbf{E}}{\partial t^2} \tag{1.7}$$

Next, use is made of the vector identity $\nabla \times \nabla \times \mathbf{E} = \nabla(\nabla \cdot \mathbf{E}) - \nabla^2 \mathbf{E}$. It is further assumed that the medium is homogeneous, meaning that ϵ and μ do not vary with position. Therefore, since $\nabla \cdot \mathbf{D} = 0$ ($\rho_v = 0$), it follows that $\nabla \cdot \mathbf{E} = 0$. Equation (1.7) becomes

$$\nabla^2 \mathbf{E} - \mu\epsilon \frac{\partial^2 \mathbf{E}}{\partial t^2} = 0 \tag{1.8}$$

Taking the curl of (1.2), substituting (1.1), and then using a similar procedure to the one that led to (1.8) result in the wave equation for the magnetic field:

$$\nabla^2 \mathbf{H} - \mu\epsilon \frac{\partial^2 \mathbf{H}}{\partial t^2} = 0 \tag{1.9}$$

The solutions of (1.8) and (1.9) will be of the form of propagating functions, which, for travel along the z axis, take the following forms: $\mathbf{E} = \mathbf{E}_0 g[t \pm (z/v)]$ and $\mathbf{H} = \mathbf{H}_0 g[t \pm (z/v)]$, where v is the velocity of wave propagation and g is any function. These results can be verified by direct substitution into (1.8) or (1.9) as appropriate. Choosing the minus sign in the argument of g means that as t increases, an increase in z is necessary to maintain a fixed value of g. Consequently, the function g (and hence the fields) will move in the positive z direction. By similar reasoning, propagation in the $-z$ direction occurs if the plus sign is chosen. Substituting the above forms of $\mathbf{E}$ or $\mathbf{H}$ into the appropriate wave equation, and assuming variation with z and t only, leads to the identification of the wave velocity as $v = 1/\sqrt{\mu\epsilon}$.

Of particular interest is the time-harmonic case, in which $g(t) = \cos(\omega t)$, leading to a wave of the form: $g[t \pm (z/v)] = \cos[\omega(t \pm z/v)] = \cos(\omega t \pm \beta z)$. The radian frequency, ω, is equal to $2\pi f$, where f is in hertz (Hz). The *propagation constant*, defined as $\beta \equiv \omega/v$, is the phase shift per unit distance of the sinusoidal wave measured along the z axis. Since the function is sinusoidal, the wave velocity is known as the *phase velocity*, v. Furthermore, with t fixed, the function will go through one complete cycle in z when $\beta z = 2\pi$. With this condition met, the distance z is equal to one *wavelength*, λ, and so $\beta = 2\pi/\lambda$.

Allowing for an overall phase, ϕ, the electric field of a wave that propagates along the z axis in a lossless medium is expressed in *real instantaneous* form as

$$\mathscr{E} = \mathbf{E}_0 \cos(\omega t \pm \beta z + \phi) = \tfrac{1}{2}\mathbf{E}_0 \exp[j(\omega t \pm \beta z + \phi)] + \text{c.c.} \tag{1.10}$$

where $\mathbf{E}_0$ represents the electric field magnitude and direction, and c.c. denotes the complex conjugate of the preceding term. The field in *complex instantaneous* form is expressed as

$$\mathbf{E}_c = \mathbf{E}_0 \exp[j(\omega t \pm \beta z + \phi)] \tag{1.11}$$

Since ω is included in β, (1.11) would adequately describe the wave if the time dependence, $\exp(j\omega t)$, were not included, which is equivalent to freezing the time at $t = 0$. The resulting *phasor form* of the field becomes a useful simplified way of expressing a single wave or a combination of waves that are all at a single frequency, ω:

$$\mathbf{E}_s = \mathbf{E}_0 \exp(\pm j\beta z + j\phi) \tag{1.12}$$

The magnetic field will have the same propagation constant and frequency and can thus be expressed in the same way as the electric field in (1.11) and (1.12). The relations between (1.10) through (1.12) are

$$\mathscr{E} = \mathrm{Re}\{\mathbf{E}_c\} = \mathrm{Re}\{\mathbf{E}_s \exp(j\omega t)\} = \tfrac{1}{2}\mathbf{E}_s \exp(j\omega t) + \mathrm{c.c.} \tag{1.13}$$

For the magnetic field, similar relations apply:

$$\mathscr{H} = \mathrm{Re}\{\mathbf{H}_c\} = \mathrm{Re}\{\mathbf{H}_s \exp(j\omega t)\} = \tfrac{1}{2}\mathbf{H}_s \exp(j\omega t) + \mathrm{c.c.} \tag{1.14}$$

Assuming the medium is sourceless and isotropic, (1.13) and (1.14) are substituted into Maxwell's equations to yield the following modified versions in terms of the phasor forms:

$$\nabla \times \mathbf{E}_s = -j\omega\mu\mathbf{H}_s \tag{1.15}$$

$$\nabla \times \mathbf{H}_s = j\omega\epsilon\mathbf{E}_s \tag{1.16}$$

$$\nabla \cdot \mathbf{D}_s = 0 \tag{1.17}$$

$$\nabla \cdot \mathbf{B}_s = 0 \tag{1.18}$$

where the time derivative of $\mathbf{E}_c$ or $\mathbf{H}_c$ results in a multiplicative factor of $j\omega$. Since the curl operation produces a vector field that is perpendicular to the one being operated on, it can be seen that $\mathbf{E}_s$ and $\mathbf{H}_s$ will always be perpendicular to each other. From here on, the subscript s will be dropped, and the symbols $\mathbf{E}$ and $\mathbf{H}$ will denote electric and magnetic field phasors.

Defining $k \equiv \omega\sqrt{\mu\epsilon}$, the wave equations (1.8) and (1.9) become

$$\nabla^2\mathbf{E} + k^2\mathbf{E} = 0 \tag{1.19}$$

$$\nabla^2\mathbf{H} + k^2\mathbf{H} = 0 \tag{1.20}$$

These two equations are known as the vector Helmholtz equations. They form the starting point for the analysis of all types of waveguides that are constructed from linear, homogeneous, and isotropic materials.

As an example, consider a wave that propagates in the positive z direction in a lossless material. Suppose that the electric and magnetic fields lie in the plane transverse to the propagation direction (xy), and furthermore, that they exhibit no variation in any direction within the transverse plane. Such a wave is an example of a *uniform plane wave*. Under these conditions, (1.19) becomes $d^2\mathbf{E}/dz^2 + k^2\mathbf{E} = 0$, whose solution is $\mathbf{E} = \mathbf{E}_0 \exp(-jkz)$. Thus in the uniform plane wave case, $\beta = k = \omega\sqrt{\mu\epsilon}$.

As another example, consider a lossless waveguide that is oriented along the z axis. The effect of the guide is to restrict the fields in the transverse (xy) plane; derivatives with respect to x or y are thus nonzero. Consider a forward z-propagating wave, whose variation with z is described by $\exp(-j\beta z)$. The phasor form of the electric field can be written as $\mathbf{E} = \mathbf{E}(x, y, z) = \mathbf{E}_0(x, y) \exp(-j\beta z)$. A similar expression can be written for $\mathbf{H}$. Substituting $\mathbf{E}$ into (1.19) results in

$$\nabla_t^2 \mathbf{E}_0 + (k^2 - \beta^2)\mathbf{E}_0 = 0 \qquad (1.21)$$

where the transverse Laplacian is $\nabla_t^2 \equiv \partial^2/\partial x^2 + \partial^2/\partial y^2$. It is seen that $\beta \neq k$ if the fields vary in the transverse direction to propagation. β will thus depend on the transverse field variation, which in turn is determined by the structure of the guide.

Consider again a uniform plane wave that propagates in the positive z direction. Suppose the electric field direction (polarization) is along x, so that its phasor form is $\mathbf{E} = E\hat{\mathbf{a}}_x = E_0 \exp(-jkz)\hat{\mathbf{a}}_x$, where $\hat{\mathbf{a}}_x$ is a unit vector along x. The magnetic field can be found in terms of $\mathbf{E}$ by using (1.15), where in this case the curl simplifies to $\nabla \times \mathbf{E} = (\partial E/\partial z)\hat{\mathbf{a}}_y = -jkE\hat{\mathbf{a}}_y$. Setting this result equal to $-j\omega\mu\mathbf{H}$ as per (1.15), it is found that $\mathbf{H} = H_0 \exp(-jkz)\hat{\mathbf{a}}_y$, where $E_0/H_0 = \omega\mu/k = \sqrt{\mu/\epsilon} \equiv \eta$. η is called the *intrinsic impedance* of the medium and is expressed in units of ohms (Ω). In free space, it becomes $\eta_0 = \sqrt{\mu_0/\epsilon_0} \approx 377\,\Omega$.

1.2. POWER FLOW

The propagating fields as described in the last section constitute the transport of energy. The power per unit cross-sectional area associated with the fields is given by the *Poynting vector*,

$$\mathbf{S} = \mathscr{E} \times \mathscr{H} \qquad (1.22)$$

in watts/m^2 (W/m^2). This result originates from Poynting's theorem, which relates the flow of power into or out of a closed system to the rate of stored energy change and power dissipation in the system. The reader is referred to any standard electromagnetics text (such as ref. 1) for the derivation and discussion of this theorem.

The electric and magnetic fields appearing in (1.22) are the real instantaneous forms, thus leading to the instantaneous power density. In practice, the *time-average* power is usually measured. This is because most detection equipment (particularly at higher frequencies) cannot respond fast enough to follow the oscillating fields, so they effectively integrate the instantaneous power. The time-average power density is given by

$$\langle \mathbf{S} \rangle = \frac{1}{T} \int_0^T \mathscr{E} \times \mathscr{H} \, dt \tag{1.23}$$

where $T = 2\pi/\omega$ is the oscillation period. Substituting (1.13) and (1.14) into (1.23) yields an expression for $\langle \mathbf{S} \rangle$ that can also be obtained through the following computation that uses the phasor forms of the fields:

$$\langle \mathbf{S} \rangle = \tfrac{1}{2} \operatorname{Re} \{ \mathbf{E} \times \mathbf{H}^* \} \tag{1.24}$$

As an example, for a uniform plane wave where $\mathbf{E} = E_0 \exp(-jkz)\hat{\mathbf{a}}_x$ and where $\mathbf{H} = (E_0/\eta) \exp(-jkz)\hat{\mathbf{a}}_y$, it is found from (1.24) that

$$\langle \mathbf{S} \rangle = \frac{|E_0|^2}{2|\eta|} \cos(\phi)\hat{\mathbf{a}}_z \quad \text{W/m}^2 \tag{1.25}$$

This allows for the possibility of a complex impedance, $\eta = |\eta| \exp(j\phi)$, which would arise when ϵ is complex. The latter occurs when the medium is conductive or when the frequency is in the vicinity of material resonances, as will be considered in Section 1.5.

As another example, consider the electric field associated with two waves in free space that propagate in opposite directions along the z axis. Suppose both waves have equal field amplitudes and polarization directions. Assuming x polarization, the total field in phasor form will be

$$\mathbf{E}_T = E_0[\exp(-jkz) + \exp(+jkz)]\hat{\mathbf{a}}_x = 2E_0 \cos(kz)\hat{\mathbf{a}}_x \tag{1.26}$$

Using (1.13), the real instantaneous form of this field will be

$$\mathscr{E}_T = 2E_0 \cos(kz) \cos(\omega t)\hat{\mathbf{a}}_x \tag{1.27}$$

This field is that of a *standing wave*, in which the spatial and temporal dependences are in two separate functions whose product yields the wavefunction. This is to be distinguished from a propagating wave, as exemplified in (1.10), where the argument of the cosine, $\omega t \pm \beta z$, dictates that the field must move along z as time changes. Thus, in a propagating wave, every position along z experiences the maximum amplitude, E_0, at some point in time. In contrast, the amplitude of (1.27), $2E_0 \cos(kz)$, is limited by its z dependence, reaching zero for all time, for example, when $kz = (2n + 1)\pi/2$.

The phasor magnetic field associated with (1.26) will be

$$\mathbf{H}_T = \frac{E_0}{\eta_0}[\exp(-jkz) - \exp(+jkz)]\hat{\mathbf{a}}_y = -j\frac{2E_0}{\eta_0}\sin(kz)\hat{\mathbf{a}}_y \qquad (1.28)$$

Using (1.26) and (1.28) in (1.24), it is found that the average power transmitted in a standing wave is zero, even though the individual powers associated with each of the propagating waves that combine to produce it are not. Another way to visualize this is to note that the Poynting vectors in the two counterpropagating waves cancel, resulting in no net flow of power. The imaginary term resulting from the product $\mathbf{E} \times \mathbf{H}^*$ is associated with the conversion of stored energy back and forth between the electric and magnetic fields at a given position z and is termed *reactive* power. Some form of standing wave appears as a result of wave reflection, as will be seen in later sections.

1.3. GROUP VELOCITY

The phase velocity of a wave is necessarily frequency dependent, since the dielectric constant of the medium and the field configuration in a waveguide are in general functions of frequency. This fact has important implications when considering the propagation of waves that are composed of more than one frequency, as would be the case, for example, in an amplitude- or frequency-modulated carrier wave. The velocity of the modulation will in general differ from that of the carrier.

To illustrate, consider a uniform plane wave composed of two equal-amplitude fields at two frequencies, ω_1 and ω_2, where it is assumed that $\omega_2 > \omega_1$. The total real instantaneous field is constructed of the sum of the two component fields:

$$\begin{aligned}
\mathcal{E}_T &= \text{Re}\{E_0[\exp(-j\beta_1 z)\exp(j\omega_1 t) + \exp(-j\beta_2 z)\exp(j\omega_2 t)]\} \\
&= 2E_0\cos(\Delta\omega t - \Delta\beta z)\cos(\omega_0 t - \beta_0 z)
\end{aligned} \qquad (1.29)$$

where $\omega_0 = \omega_1 + \Delta\omega = \omega_2 - \Delta\omega$ and $\beta_0 = \beta_1 + \Delta\beta = \beta_2 - \Delta\beta$.

Equation (1.29) represents a carrier wave at frequency ω_0 that is modulated by a sinusoidal envelope at the "beat" frequency $\Delta\omega$, as depicted in Fig. 1.1.

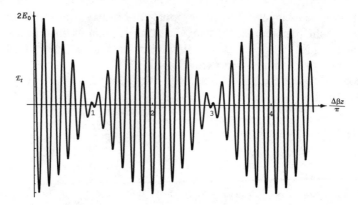

Figure 1.1. z Dependence of the electric field of (1.29) at $t = 0$, where $\omega_0/\Delta\omega = 24$.

The carrier phase velocity is $v_{pc} = \omega_0/\beta_0$; the envelope velocity, $v_{pe} = \Delta\omega/\Delta\beta$, will be equal to that of the carrier, provided $\beta(\omega)$ and ω are related to each other through a multiplying factor that does not vary with frequency. For example, suppose the permeability is μ_0, but the refractive index is frequency dependent. Thus $\beta(\omega) = k(\omega) = n^\omega \omega/c$. The carrier and envelope velocities are

$$v_{pc} = \frac{\omega_0}{\beta_0} = \frac{c}{n^{\omega_0}} \tag{1.30}$$

$$v_{pe} = \frac{\Delta\omega}{\Delta\beta} = \frac{(\omega_2 - \omega_1)c}{\omega_2 n^{\omega_2} - \omega_1 n^{\omega_1}} \tag{1.31}$$

Normally, $n^{\omega_2} > n^{\omega_1}$, as will be shown in Section 1.5. Therefore the envelope velocity is generally less than that of the carrier.

As $\Delta\omega$ becomes small, the quantity $\Delta\omega/\Delta\beta$ approaches $d\omega/d\beta$, or the slope of the ω versus β curve for a given medium, as shown in Fig. 1.2. $d\omega/d\beta$ is the frequency-dependent *group velocity* function, $v_g(\omega)$, for the medium. It is interpreted as the propagation velocity of a "group" of waves whose frequencies are distributed over an infinitesimally small bandwidth, centered on frequency ω. In most situations, it has the additional interpretation as the velocity of energy flow in the wave. The small bandwidth requirement in the definition arises since, if true, the ω versus β curve will appear straight over the small frequency range, and thus v_g will be sharply defined. Alternately, with a broad frequency spectrum, the slope of the ω versus β curve may exhibit changes over the range of the spectrum. As a result, different components of the spectrum will propagate at different group velocities, leading to signal distortion. The details of this latter problem, known as *group velocity dispersion*, will be addressed in Chapter 5.

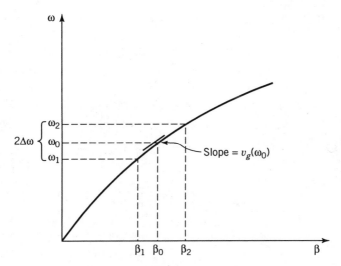

Figure 1.2. ω–β plot for a typical material, showing the variation of group velocity with frequency.

1.4. REFLECTION AND TRANSMISSION OF WAVES AT PLANE INTERFACES

The phenomena of reflection and transmission of plane waves at interfaces between dielectrics are useful in the interpretion of waveguide behavior. In this section, the governing equations for plane interfaces are derived for the two fundamental wave polarization cases; these are diagramed in Fig. 1.3. In both cases, the interface lies in the yz plane. The two media have refractive indices n_1 ($x < 0$) and n_2 ($x > 0$); both are assumed to have permeability μ_0. A single plane wave is incident from the region above the interface at angle θ_1 from the x axis. The propagation direction and frequency of the incident wave are characterized by a *wavevector*, $\mathbf{k}_1 = n_1 k_0 [\cos\theta_1 \hat{\mathbf{a}}_x + \sin\theta_1 \hat{\mathbf{a}}_z]$, where $k_0 \equiv \omega\sqrt{\mu_0\epsilon_0} = \omega/c$, and where it is seen that $|\mathbf{k}_1| = (\mathbf{k}_1 \cdot \mathbf{k}_1)^{1/2} = n_1 k_0$.

The direction of $\mathbf{k}_1$ is normal to the surfaces of constant phase. This means that the phase shift per unit distance is maximized in the direction of $\mathbf{k}_1$; the wave phase velocity is thus minimized along this direction. Assuming isotropic media, $\mathbf{k}$ also has the interpretation of the direction of power flow in the wave. The latter quantity is expressed through the time-average Poynting vector (1.24).

The *plane of incidence* is defined as the plane spanned by the incident wavevector and the normal vector to the surface (in this case the x axis). The two cases in Fig. 1.3 differ in the direction assigned to the electric field vector of the incident wave. In the left figure, the field is polarized in the direction perpendicular to the plane of incidence. $\mathbf{E}$ thus lies in a plane that is perpendicular to the z axis and so is alternately referred to as *transverse electric* (or

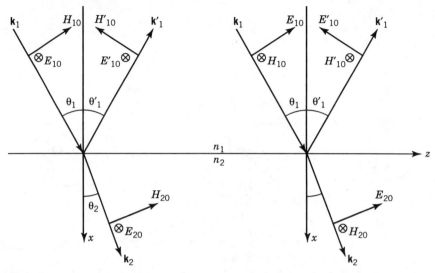

Figure 1.3. Reflection and transmission geometries for plane waves exhibiting TE polarization (left) and TM polarization (right).

TE) polarization. In the right figure, the electric field lies *in* the plane of incidence or is parallel polarized. Since the magnetic field will now lie in the plane perpendicular to the z axis, this case is also referred to as *transverse magnetic* (or TM) polarization. Any other polarization direction can be expressed as a combination of these two cases.†

It is desired to determine the magnitudes of the reflected and transmitted electric fields, as well as the reflection and refraction angles, θ_1' and θ_2. To accomplish this, use is made of the boundary conditions that require continuity of tangential electric and magnetic fields across the interface:

$$E_{t1}|_{x=0} + E_{t1}'|_{x=0} = E_{t2}|_{x=0} \tag{1.32}$$

$$H_{t1}|_{x=0} + H_{t1}'|_{x=0} = H_{t2}|_{x=0} \tag{1.33}$$

TE polarization is considered first. The incident, reflected, and transmitted electric field magnitudes take the form

$$E_{y1} = E_{10} \exp(-j\mathbf{k}_1 \cdot \mathbf{r}) \tag{1.34}$$

---

† The TE and TM polarizations are sometimes called s and p polarizations, respectively; these originate from the German *senkrecht* (perpendicular) and *parallel* (parallel), describing the electric field as it relates to the plane of incidence.

$$E'_{y1} = E'_{10} \exp(-j\mathbf{k}'_1 \cdot \mathbf{r}) \tag{1.35}$$

$$E_{y2} = E_{20} \exp(-j\mathbf{k}_2 \cdot \mathbf{r}) \tag{1.36}$$

where $\mathbf{r} = x\hat{\mathbf{a}}_x + y\hat{\mathbf{a}}_y + z\hat{\mathbf{a}}_z$. The wavevectors for the TE and TM cases are expressed as follows:

$$\mathbf{k}_1 = k_{x1}\hat{\mathbf{a}}_x + k_{z1}\hat{\mathbf{a}}_z = n_1 k_0 \cos(\theta_1)\hat{\mathbf{a}}_x + n_1 k_0 \sin(\theta_1)\hat{\mathbf{a}}_z \tag{1.37}$$
$$\mathbf{k}'_1 = -k'_{x1}\hat{\mathbf{a}}_x + k'_{z1}\hat{\mathbf{a}}_z = -n_1 k_0 \cos(\theta'_1)\hat{\mathbf{a}}_x + n_1 k_0 \sin(\theta'_1)\hat{\mathbf{a}}_z \tag{1.38}$$
$$\mathbf{k}_2 = k_{x2}\hat{\mathbf{a}}_x + k_{z2}\hat{\mathbf{a}}_z = n_2 k_0 \cos(\theta_2)\hat{\mathbf{a}}_x + n_2 k_0 \sin(\theta_2)\hat{\mathbf{a}}_z \tag{1.39}$$

Note that in order to satisfy the boundary conditions (1.32) and (1.33) for all points on the interface, all three waves must exhibit the same phase shift per unit distance along the interface. Therefore it must be true that $k_{z1} = k'_{z1} = k_{z2}$. With increasing time, this condition translates to that of requiring equal phase velocities for the three waves along the interface, where $v_{pz} = \omega/k_{z1}$ for the incident wave. As a consequence, it follows from (1.37) and (1.38) that $\theta_1 = \theta'_1$ (the incidence angle equals the reflection angle); also, from (1.37) and (1.38) it is found that

$$n_1 \sin \theta_1 = n_2 \sin \theta_2 \tag{1.40}$$

The above equation, relating the incident and transmitted angles, is known as *Snell's law of refraction*.

For TE waves, (1.32) and (1.33) can be expressed in terms of the electric and magnetic field amplitudes as follows:

$$E_{10} + E'_{10} = E_{20} \tag{1.41}$$

$$H_{10} \cos \theta_1 - H'_{10} \cos \theta_1 = H_{20} \cos \theta_2 \tag{1.42}$$

The ratios of the amplitudes will be $E_{10}/H_{10} = E'_{10}/H'_{10} = \eta_1 = \eta_0/n_1$ and $E_{20}/H_{20} = \eta_0/n_2$. Using these ratios, (1.42) becomes

$$E_{10}n_1 \cos \theta_1 - E'_{10}n_1 \cos \theta_1 = E_{20}n_2 \cos \theta_2 \tag{1.43}$$

Solving (1.42) and (1.43) together leads to expressions for the ratios of reflected-to-incident and transmitted-to-incident fields (reflection and transmission coefficients). These are

$$\Gamma_{TE} \equiv \frac{E'_{10}}{E_{10}} = \frac{n_1 \cos\theta_1 - n_2\cos\theta_2}{n_1\cos\theta_1 + n_2\cos\theta_2} = \frac{k_{x1} - k_{x2}}{k_{x1} + k_{x2}} \qquad (1.44)$$

$$T_{TE} \equiv \frac{E_{20}}{E_{10}} = \frac{2k_{x1}}{k_{x1} + k_{x2}} = 1 + \Gamma_{TE} \qquad (1.45)$$

A similar procedure is used to analyze TM reflection (Problem 1.3). Resulting are reflection and transmission coefficients, given as

$$\Gamma_{TM} \equiv \frac{E'_{10}}{E_{10}} = \frac{\cos\theta_1/n_1 - \cos\theta_2/n_2}{\cos\theta_1/n_1 + \cos\theta_2/n_2} = \frac{n_2^2 k_{x1} - n_1^2 k_{x2}}{n_2^2 k_{x1} + n_1^2 k_{x2}} \qquad (1.46)$$

$$T_{TM} \equiv \frac{E_{20}}{E_{10}} = \frac{2n_2^2 k_{x1}}{n_2^2 k_{x1} + n_1^2 k_{x2}} = 1 + \Gamma_{TM} \qquad (1.47)$$

Equations (1.44) through (1.47) are known as the *Fresnel equations*, describing reflection and transmission of plane waves at plane interfaces in nonmagnetic materials.

A few special cases are of interest. It is possible to achieve zero reflection at a certain angle for TM polarization. This is shown by setting (1.46) equal to zero, to obtain the condition

$$n_2 \cos\theta_1 = n_1 \cos\theta_2 \qquad (1.48)$$

Using Snell's law (1.40) and solving for the incident angle result in

$$\cos\theta_1 = \cos\theta_B = \left(\frac{n_1^2}{n_1^2 + n_2^2}\right)^{1/2} \qquad (1.49)$$

θ_B is known as the *Brewster angle*, at which total transmission is achieved for TM polarization. For TE waves, one would attempt to find an incident angle at which (1.44) would be zero. No solution exists in that case, leading to the additional interpretation of θ_B as the "polarizing angle" at which the reflected portion of incident light of any polarization will be TE polarized only.

Another special case is that of total reflection, which can occur for either polarization. Snell's law can be expressed as

$$\cos\theta_2 = \left(1 - \frac{n_1^2}{n_2^2}\sin^2\theta_1\right)^{1/2} \qquad (1.50)$$

Note that $\cos\theta_2 = 0$ if $\sin\theta_1 = n_2/n_1$, in which case k_{x2} (given by $n_2 k_0 \cos\theta_2$) is zero. Furthermore, if $\sin\theta_1 > n_2/n_1$, then k_{x2} will be imaginary. With either of the above conditions met, the power reflection coefficients, given by $\Gamma_{TE}\Gamma_{TE}^*$ and $\Gamma_{TM}\Gamma_{TM}^*$ will be unity. The condition of total reflection for either polarization is therefore

$$\theta_1 \geq \theta_c = \sin^{-1}(n_2/n_1) \tag{1.51}$$

where θ_c is the *critical angle* of total reflection. It is apparent that for total reflection to occur, n_2 must be less than n_1.

1.5. MATERIAL RESONANCES AND THEIR EFFECTS ON WAVE PROPAGATION

Material parameters such as refractive index, dispersion, and absorption have their fundamental origins in atomic and molecular resonances that are present in the medium. In classical terms, an electron or an entire atom can be set into a state of oscillation when displaced from its equilibrium position. The restoring forces arise from adjacent molecules or, as in the case of a single electron, from the nucleus of a single atom. The resonant frequency of the oscillation is determined by the magnitude of the restoring forces and the mass of the oscillating body. The time-varying electric field of the incident light provides the driving force on the oscillator. Since oscillation involves periodic charge displacement, a dipole moment is produced that varies sinusoidally in time with the oscillation. This process, along with the subsequent interaction of all the induced dipole moments with the optical field, forms the basis of the classical model of light interaction with matter. Although this model possesses significant limitations in some situations, it is nevertheless very accurate in quantifying absorption and dispersion in many materials.

Consider a simple material composed of an ensemble of identical electron oscillators. These lie along the propagation path of a wave characterized by complex field, $\mathbf{E}_c$, of the form given by (1.11). The displacement of the electron in a specific oscillator at position z amounts to a shifting or distortion of the electron cloud that surrounds the nucleus of the atom in question. Figure 1.4a shows the cloud before and after displacement and indicates pictorially the formation of a dipole moment. The Coulomb force that resists displacement is modeled as that of a classical spring, having force constant k_s. Figure 1.4b shows the spring model, in which the electron, having mass m and charge e (taken as positive), is considered localized at displaced position $\mathbf{Q}_c$. The restoring force is expressed as $\mathbf{F}_r = -k_s\mathbf{Q}_c$.

The displacing force is provided by the optical electric field, $\mathbf{E}_c$, which oscillates at frequency ω; this force is given by $\mathbf{F}_a = -e\mathbf{E}_c$. A damping force, proportional to the electron velocity, $\mathbf{v}$, is also present and is given as $\mathbf{F}_d = -m\zeta\mathbf{v}$. The damping coefficient, ζ, is used to model effects such as spontaneous emission and the influences of neighboring atoms.

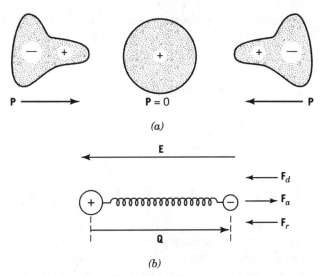

Figure 1.4. An electron oscillator (a) and its spring model representation (b).

The force equation for the electron takes the form

$$m \frac{d^2\mathbf{Q}_c}{dt^2} + m\zeta \frac{d\mathbf{Q}_c}{dt} + k_s\mathbf{Q}_c = -e\mathbf{E} \tag{1.52}$$

where $\mathbf{E}_c$ is given by (1.11). The displacement is thus assumed to be of the form

$$\mathbf{Q}_c = \mathbf{Q}_0 \exp(-j\beta z) \exp(j\omega t) \tag{1.52a}$$

Substituting (1.11) and (1.52a) into (1.52) and noting that time differentiation yields a factor of $j\omega$, we obtain

$$\mathbf{Q}_c = \frac{-e\mathbf{E}_c/m}{(\omega_0^2 - \omega^2) + j\omega\zeta} \tag{1.53}$$

where the resonant frequency $\omega_0 = \sqrt{k_s/m}$. The dipole moment for the single electron is now given by $\mathbf{p}_c = -e\mathbf{Q}_c$. The net dipole moment within a small volume Δv, located at position z, can be found by summing the individual dipole moments within that volume. The complex polarization field at z is then found by dividing the summation by Δv and taking the limit as Δv shrinks to a volume, δ, that is small enough to be considered a point, yet large enough to enclose several dipoles:

$$\mathbf{P}_c = \lim_{\Delta v \to \delta} \sum_{i=1}^{N \Delta v} \mathbf{p}_{ci} = N\mathbf{p}_c \quad \text{C/m}^2 \tag{1.54}$$

where N is the number of oscillators per unit volume, and where the second equality in (1.54) applies to the special case in which all oscillators are identical. By using (1.53), the polarization becomes

$$\mathbf{P}_c = \left(\frac{Ne^2/m}{(\omega_0^2 - \omega^2) + j\omega\zeta} \right) \mathbf{E}_c = \epsilon_0 \chi_e \mathbf{E}_c \tag{1.55}$$

where the dimensionless function, χ_e, is the linear electric susceptibility of the material.† Solving for χ_e and separating real and imaginary parts, we obtain

$$\chi_e = \chi_e' - j\chi_e'' = \frac{Ne^2}{\epsilon_0 m} \frac{(\omega_0^2 - \omega^2) - j\omega\zeta}{(\omega_0^2 - \omega^2)^2 + \omega^2\zeta^2} \tag{1.56}$$

A simplified form of (1.56) can be obtained by assuming near-resonance operation, such that $\omega \approx \omega_0$. Under this approximation, (1.56) becomes

$$\chi_e' - j\chi_e'' \approx -\frac{Ne^2}{\epsilon_0 m \omega_0 \zeta} \left(\frac{j + \delta}{1 + \delta^2} \right) \tag{1.57}$$

where $\delta \equiv (2/\zeta)(\omega - \omega_0)$ is the normalized detuning parameter. The *linewidth* of the resonance is the frequency offset from ω_0 at which $\delta = 1$, such that the imaginary part of (1.57) is reduced to one-half of its peak value. This occurs when $|\omega - \omega_0| = \zeta/2$. The damping coefficient, ζ, is thus the full width in frequency at the half-maximum points of χ_e''.

Equation (1.57) is plotted in Fig. 1.5. The shape of the curve for χ_e'' is Lorentzian, having a full width at one-half its peak value of ζ (the half-maximum points occur at $\delta = \pm 1$). The peaks of χ_e' have values of plus or minus one-half the peak value of χ_e'', and they coincide in frequency with the half-maximum positions of χ_e'' as can readily be shown (Problem 1.9). Again, the plot is made under the assumption of near-resonance, which further implies that ζ must be sufficiently small to allow reasonable accuracy over the indicated range of frequencies.

The effects of the complex susceptibility on a propagating wave are illustrated by considering a uniform plane wave that propagates in the positive z

---

† Note that with a complex susceptibility, the real instantaneous polarization field will be $\mathscr{P} = \text{Re}\{\epsilon_0 \chi_e \mathbf{E}_c \exp(j\omega t)\}$.

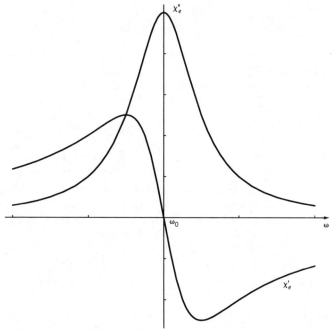

Figure 1.5. Plots of the real and imaginary parts of the spring model susceptibility under the near-resonance approximation (1.57). Note that the imaginary part is plotted as positive.

direction. The phasor form of the wave magnitude will be

$$E = E_0 \exp(-jkz) = E_0 \exp(-\alpha z) \exp(-j\beta z) \qquad (1.58)$$

where α and β, the real and imaginary parts of jk, are the attenuation and propagation constants, respectively. k is related to the complex refractive index through $k = nk_0$, where

$$n = (1 + \chi'_e - j\chi''_e)^{1/2} = n' - jn'' \qquad (1.59)$$

The real and imaginary parts of n (index and extinction coefficient) are found to be

$$n' = \sqrt{\frac{1 + \chi'_e}{2}} \left(\sqrt{1 + \frac{(\chi''_e)^2}{(1 + \chi'_e)^2}} + 1 \right)^{1/2} \qquad (1.60)$$

$$n'' = \sqrt{\frac{1+\chi_e'}{2}} \left(\sqrt{1 + \frac{(\chi_e'')^2}{(1+\chi_e')^2}} - 1 \right)^{1/2} \tag{1.61}$$

The propagation constant is $\beta = n'k_0$ and the attenuation coefficient is $\alpha = n''k_0$.

A special case is that in which $\chi_e'' \ll 1 + \chi_e'$. The real and imaginary parts of n become $n' \approx (1+\chi_e')^{1/2}$ and $n'' \approx (\chi_e''/2)(1+\chi_e')^{-1/2}$. The above behavior illustrates the reason that χ_e'' and χ_e' are referred to, respectively, as the absorptive and refractive index parts of the susceptibility. This case corresponds to that of low losses, which would occur as a result of a weak resonance, or as the operating frequency is tuned far away from resonance.

A further simplification arises at frequencies far from resonance. In such cases, (1.56) can be modified under the assumption that the damping coefficient, ζ, is reduced to zero. The reduction of damping can be explained in a simplistic way by noting that, far from resonance, electron displacements and the resulting influence of surrounding material will be small. Consequently, $\chi_e'' \approx 0$ and $\chi_e' \approx (Ne^2/\epsilon_0 m)(\omega_0^2 - \omega^2)^{-1}$. It is often convenient to express the latter equation in terms of wavelength:

$$\chi_e' \approx \left[\frac{Ne^2\lambda_0^2}{(2\pi c)^2\epsilon_0 m} \right] \frac{\lambda^2}{\lambda^2 - \lambda_0^2} = \frac{A\lambda^2}{\lambda^2 - \lambda_0^2} \tag{1.62}$$

where λ and λ_0 are the free-space wavelengths corresponding to ω and ω_0. Since $\chi_e'' \approx 0$, we have $n \approx n' \approx (1+\chi_e')^{1/2}$. Additionally, more than one resonance in the material can be accounted for by summing the refractive index contributions from each resonance. The result is the *Sellmeier formula* for refractive index, which is commonly expressed in the form

$$n^2 - 1 = \sum_i \frac{A_i\lambda^2}{\lambda^2 - \lambda_{0i}^2} \tag{1.63}$$

where A_i and λ_{0i} are, respectively, the magnitude and resonant wavelength of the ith resonance.

It will be noted that in regions of the spectrum away from any resonance, the real part of the susceptibility (related to refractive index) will decrease with increasing wavelength. Another feature is that in the frequency region between two resonances, the curve for χ_e' will exhibit a point of inflection, as shown in Fig. 1.6. The occurrence of an inflection point means that the material exhibits *zero group velocity dispersion* at that frequency. This principle will be discussed in detail in Chapter 5.

The above discussion has presented formulations for material properties in frequency domain. Specifically, whereas in isotropic media $\mathbf{P}_c(\omega) =$

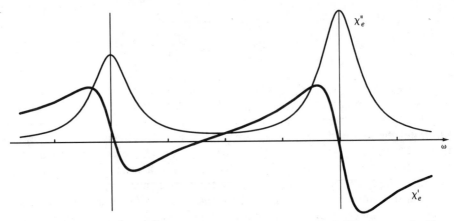

Figure 1.6. Plots of χ_e' and χ_e'' for a material having two resonances of different strengths. Note the inflection point in the χ_e' curve, which occurs at the zero dispersion frequency.

$\epsilon_0\chi_e(\omega)\mathbf{E}_c(\omega)$ is written as the simple product of two frequency-dependent functions, the general relation between electric field and polarization in time domain is expressed through the convolution

$$\mathbf{P}_c(t) = \int_{-\infty}^{t} \chi_e(t - \tau)\mathbf{E}_c(\tau)d\tau \qquad (1.64)$$

where the impulse response, $\chi_e(t)$, is the Fourier transform of $\chi_e(\omega)$. In most of the treatment throughout this book, off-resonance operation is assumed. One consequence of this is that dielectric response times in most materials can be treated as instantaneous, so that the susceptibility will exhibit a delta function time dependence: $\chi_e(t) \approx \chi_{e0}\delta(t)$. Thus $\mathbf{P}_c(t) \approx \chi_{e0}\mathbf{E}_c(t)$. This simplification is of particular importance when analyzing the propagation of optical pulses, as will be described in Chapters 5 and 8. Aside from the advantage of a simplified analysis, the occurrence of some important nonlinear effects, such as the formation of optical solitons (to be considered in Chapter 8), is possible only in media with response times that are much shorter than the pulse durations.

PROBLEMS

1.1. A uniform plane wave propagates in the positive z direction in a medium in which $\epsilon_r = 4.0$ and $\mu_r = 1.0$. The wave is sinusoidal with frequency $f = 5$ MHz and amplitude A.

   a) Calculate the wavelength and phase velocity.

   b) Write down the expressions for the electric and magnetic fields in phasor, complex instantaneous, and real instantaneous forms. Take the electric field to be polarized along x.

c) Determine the power density of the wave in W/m^2, in terms of the electric field amplitude.

1.2. An important part of the study of dielectric waveguides is the evanescent wave, existing outside the guiding layer. A typical electric field for such a wave in free space is given in real instantaneous form as

$$\mathscr{E} = E_0 \exp(-\alpha x) \cos(\omega t - \beta z)\hat{a}_y \qquad x \geq 0$$

a) Express the field in phasor form and indicate its direction of propagation, wavelength, and phase velocity.

b) Using Eq. (1.15), find the associated magnetic field vector in phasor form (note that there will be two components).

c) Assuming a sourceless medium, determine α as a function of β, ω, μ_0, and ϵ_0, such that all of Maxwell's equations are satisfied by the two fields.

1.3. Derive the Fresnel equations, (1.43) and (1.44), that determine reflectivity and transmissivity for TM polarized light.

1.4. Show that no Brewster angle exists for TE polarization.

1.5. Determine the sum of the incident and refracted angles for Brewster angle incidence.

1.6. Show that the fraction of the incident *power* in a plane wave reflected from a dielectric interface is $|\Gamma|^2$, and thus the fraction of power transmitted through the interface is $1 - |\Gamma|^2$.

1.7. Light having random polarization (equal TE and TM components) is incident onto an interface between two media at the Brewster angle. Refractive indices are n_1 (incident medium) and n_2. Determine an expression for the percentage of the incident power that is reflected.

1.8. Assuming propagation in a medium having complex intrinsic impedance, η_c, determine the magnetic field phasor associated with the electric field of (1.57). Show that the ratio of the average powers at positions z and $z = 0$ will be

$$\frac{P(z)}{P(0)} = \exp(-2\alpha z) = 10^{-0.1\alpha_p z}$$

where $\alpha_p = 8.69\alpha$. α_p is the power attenuation coefficient in decibels (dB) per unit length.

1.9. Show that (1.55) reduces to (1.56) under the near-resonance approximation ($\omega \approx \omega_0$). Also determine the peak values of χ_e'' and their locations with respect to ω_0.

1.10.   Consider two waves having different frequencies, ω_1 and ω_2, that propagate in opposite directions along the z axis. Show that the resulting "standing wave" field distribution moves along the z axis at a velocity proportional to the frequency difference.

REFERENCES

1. S. Ramo, J. R. Whinnery, and T. Van Duzer, *Fields and Waves in Communication Electronics*, 3rd ed. Wiley, New York, 1994.
2. F. A. Jenkins and H. E. White, *Fundamentals of Optics*. McGraw-Hill, New York, 1976.
3. E. Hecht and A. Zajac, *Optics*. Addison-Wesley, Menlo Park, CA, 1979.
4. D. K. Cheng, *Field and Wave Electromagnetics*, 2nd ed. Addison-Wesley, Reading, MA, 1989.
5. J. D. Kraus, *Electromagnetics*, 4th ed. McGraw-Hill, New York, 1991.

2

Symmetric Dielectric Slab Waveguides

As a first step in the study of optical waveguides, the light-guiding properties of the dielectric slab waveguide are presented in this chapter. This structure is of great importance in the area of integrated optics, where it is a basic component in the construction of many devices. The properties of the guide (field configuration, behavior with frequency, etc.) are qualitatively similar to the optical fiber but are simpler and more easily understood. The special case of symmetric geometry, having equal refractive indices above and below the guiding layer, is presented here. The analysis of nonsymmetric structures is similar and is covered in other texts [1,2].

The symmetric slab geometry is shown in Fig. 2.1. The guide is assumed infinite in the y direction, implying that no electromagnetic quantity will vary in this direction. Three lossless dielectric regions comprise the structure. The middle layer, known as the *slab* or *film*, is of thickness d and has refractive index n_1, where $n_1 > n_2$. The upper and lower regions, having index n_2, are semi-infinite in the positive and negative x directions. The permeability in all regions is assumed to be that of free space, μ_0. Light is launched into the slab from the left, and the resulting guided modes propagate in the positive z direction.

2.1. RAY ANALYSIS OF THE SLAB WAVEGUIDE

It is possible to analyze the waveguide modes as superpositions of plane waves that propagate in zigzag paths characterized by *rays*. This approach is easiest to implement in media having boundaries that extend over distances very large compared to a wavelength, and in which local index variations are very small over distances comparable to a wavelength. Light rays are confined within the structure by the mechanism of total reflection at the interfaces. This will occur if $\theta_1 \geq \theta_c$, where θ_1 is the ray angle measured with respect to the normal to the interface, and θ_c is the critical angle, given by $\sin \theta_c = n_2/n_1$. It is apparent that this condition can be satisfied only if $n_1 > n_2$.

Figure 2.1 shows a single ray propagating in a zigzag path, reflecting from both interfaces at angle θ_1. Associated with this ray are two plane waves. One propagates along the upward path with wavevector $\mathbf{k}_u$; the other propagates along the downward path with wavevector $\mathbf{k}_d$. The two vectors are expressed in terms of their cartesian components as $\mathbf{k}_u = k_{x1}\hat{\mathbf{a}}_x + \beta\hat{\mathbf{a}}_z$ and $\mathbf{k}_d = -k_{x1}\hat{\mathbf{a}}_x + \beta\hat{\mathbf{a}}_z$.

21

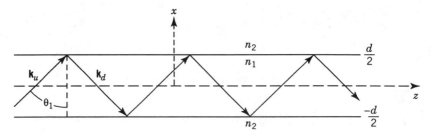

Figure 2.1. Symmetric slab ray geometry.

To achieve a propagating mode, it is necessary that the wavefronts associated with the first upward propagating ray be exactly coincident with those of all subsequent upward rays. The net upward propagating power will thus be in the form of a single plane wave. A similar requirement holds for the downward rays. If the above conditions are not satisfied, then the composite wave (or mode) will not propagate, since destructive interference would occur between the various reflected components. For a given wavelength, guide thickness, and choice of indices, only certain values of θ_1 will satisfy the requirement. Each allowed θ_1 value corresponds to a single mode of the guide.

Two distinct mode types can exist, depending on the orientation of the electric and magnetic fields of the plane waves. Figure 2.2 shows the two types. With the electric field polarized in the xz plane, the magnetic field lies entirely in the transverse plane (along the y axis), forming a transverse magnetic (or TM) mode. The opposite case occurs when **H** lies in the xz plane, leaving **E** in the transverse plane, resulting in a transverse electric (or TE) mode.

The upward and downward propagating fields are expressed in the standard phasor form, shown here for a single component of **E**:

$$E_u = E_0 \exp(-j\mathbf{k}_u \cdot \mathbf{r}) = E_0 \exp(-jk_{x1}x) \exp(-j\beta z) \qquad (2.1)$$

$$E_d = E_0 \exp(-j\mathbf{k}_d \cdot \mathbf{r}) = E_0 \exp(+jk_{x1}x) \exp(-j\beta z) \qquad (2.2)$$

where $\mathbf{r} = x\hat{\mathbf{a}}_x + z\hat{\mathbf{a}}_z$. The wavevector magnitudes are $|\mathbf{k}_u| = |\mathbf{k}_d| = k_1 = n_1 k_0$. The

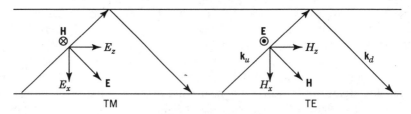

Figure 2.2. The two types of slab waveguide modes.

wavevector component magnitudes are expressed as

$$k_{x1} = k_1 \cos \theta_1 = n_1 k_0 \cos \theta_1 \qquad (2.3)$$

and

$$\beta = k_1 \sin \theta_1 = n_1 k_0 \sin \theta_1 \qquad (2.4)$$

where $k_0 \equiv \omega\sqrt{\mu_0\epsilon_0} = 2\pi/\lambda$ is the free-space propagation constant, and λ is the wavelength of light measured in vacuum. The z component of either wavevector, β (also the mode propagation constant), is one of the quantities that is necessary to determine for each mode of a guide. From β, the mode phase velocity (ω/β) and group velocity ($d\omega/d\beta$) can be found. The dependence of the latter quantity on frequency enables the evaluation of signal distortion that will occur in a given mode. It is clear that, at a given frequency, different modes will have different values of β and may consequently have different group velocities. Hence, in a multimode guide, in which an appreciable fraction of the signal power is carried in each mode, the different propagation times of the modes may produce severe signal distortion. For this reason, a guide that propagates light in a single mode is preferable.

To study the wave reflection process, the possibility of propagating fields above and below the slab must be included. To see this, consider the "leaky" wave case shown in Fig. 2.3. In it, reflection of a slab ray at each interface results in partial transmission into the upper and lower regions. The wavevectors of the transmitted rays are of magnitude k_2. Considering the upper region ($x \geq d/2$), $\mathbf{k}_2$ is projected into x and z components, k_{x2} and β, respectively. The z components of $\mathbf{k}_1$ and $\mathbf{k}_2$ must be equal to enable the field boundary conditions to be satisfied at all positions along the interface at all times. The transmitted wavevectors in the lower region ($x \leq -d/2$) are projected in a similar way, resulting in x- and z-component magnitudes that are identical to those found in the upper region.

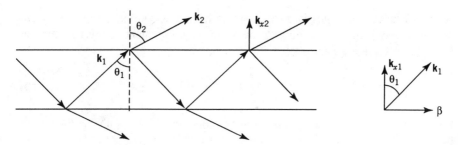

Figure 2.3. Wavevector geometries at the interfaces.

It is necessary to determine (1) the relationship between the ray angles, θ_1 and θ_2, and the refractive indices and (2) the relationship between the incident and reflected wave amplitudes. The first relationship is given by Snell's law: $n_1 \sin \theta_1 = n_2 \sin \theta_2$. The amplitude relationship is found through the Fresnel equations, (1.44) and (1.46), providing reflection coefficients:

$$\Gamma_{TE} = \frac{k_{x1} - k_{x2}}{k_{x1} + k_{x2}} = \frac{n_1 \cos \theta_1 - n_2 \cos \theta_2}{n_1 \cos \theta_1 + n_2 \cos \theta_2} \tag{2.5}$$

$$\Gamma_{TM} = \frac{n_2^2 k_{x1} - n_1^2 k_{x2}}{n_2^2 k_{x1} + n_1^2 k_{x2}} = \frac{n_2 \sec \theta_2 - n_1 \sec \theta_1}{n_2 \sec \theta_2 + n_1 \sec \theta_1} \tag{2.6}$$

To achieve a completely guided mode, the waves must be totally reflected at both interfaces; that is, the magnitudes of Γ_{TE} (or Γ_{TM}) will be unity. As explained in Chapter 1, total reflection will occur when $\theta_1 \geq \theta_c$.

An electric field component in the upper region will assume the phasor form:

$$E_2 = E_{20} \exp(-j\mathbf{k}_2 \cdot \mathbf{r}) = E_{20} \exp(-jk_{x2}x) \exp(-j\beta z) \tag{2.7}$$

When the critical angle is exceeded in the slab, k_{x2} becomes imaginary and can be expressed in terms of a real attenuation coefficient as

$$k_{x2} = -j\gamma_2 \qquad \theta_1 > \theta_c \tag{2.8}$$

E_2 now becomes $E_2 = E_{20} \exp(-\gamma_2 x) \exp(-j\beta z)$. This is the phasor expression for a *surface* or *evanescent* wave, which propagates only in the z direction and which decreases in amplitude as the magnitude of x increases in the upper and lower regions. Using Snell's law, γ_2 can be expressed as

$$\gamma_2 = jk_{x2} = jn_2 k_0 \cos \theta_2 = jn_2 k_0 (-j) \left[\left(\frac{n_1}{n_2} \right)^2 \sin^2 \theta_1 - 1 \right]^{1/2}$$

$$= (\beta^2 - n_2^2 k_0^2)^{1/2} \tag{2.9}$$

Note that large θ_1 implies large γ_2, leading to tighter confinement of the wave within the slab. Inside the slab, the wavevector components obey the relation $k_1^2 = n_1^2 k_0^2 = k_{x1}^2 + \beta^2$. Therefore

$$k_{x1} = (n_1^2 k_0^2 - \beta^2)^{1/2} \tag{2.10}$$

~mpletely guided (no radiation leakage), k_{x1} and γ_2 must both
and (2.10), we observe that the values of β for a completely
confined to the range

$$n_2 k_0 < \beta < n_1 k_0 \tag{2.11}$$

Recall that phase velocity is $v_p = \omega/\beta$ and $k_0 = \omega/c$. Using these in (2.11) results in

$$\frac{c}{n_1} < v_p < \frac{c}{n_2} \tag{2.12}$$

as the allowed range of phase velocities for a guided mode.

With the total reflection condition established, the plane wave will experience a phase shift on reflection, 2ϕ, which will be a function of θ_1. Considering the TE case with total reflection, we see that (2.5) becomes

$$\Gamma_{TE} = \frac{k_{x1} + j\gamma_2}{k_{x1} - j\gamma_2} \tag{2.13}$$

By defining $Z = k_{x1} + j\gamma_2$, (2.13) becomes $\Gamma_{TE} = Z/Z^* = \exp(j2\phi_{TE})$, where

$$\phi_{TE} = \tan^{-1}\left(\frac{\gamma_2}{k_{x1}}\right) = \tan^{-1}\left(\frac{(n_1^2 \sin^2\theta_1 - n_2^2)^{1/2}}{n_1 \cos\theta_1}\right) \tag{2.14}$$

In a similar manner, it can be shown that, for the TM case,

$$\phi_{TM} = \tan^{-1}\left(\frac{n_1^2}{n_2^2} \frac{(n_1^2 \sin^2\theta_1 - n_2^2)^{1/2}}{n_1 \cos\theta_1}\right) \tag{2.15}$$

From (2.14) and (2.15) it is observed that ϕ_{TE} and ϕ_{TM} will increase from 0 to $\pi/2$ as θ_1 increases from θ_c to $\pi/2$.

The eigenvalue equations for the propagating modes can now be derived. The requirement is that all upward propagating plane waves in the guide must be precisely in phase (the same is true for all downward propagating waves). Consider one complete "bounce" of a slab ray, consisting of two traversals of the slab and two reflections. The above condition will be satisfied if the net *transverse* phase shift of the wave over this path is an integer multiple of 2π (it is helpful to use appropriate sketches of the propagating phase fronts to illustrate this fact). Consider a starting electric field value, E_s, and its value at the end of a complete round trip, E_f. The two are related by

$$E_f = E_s \exp(-jk_{x1}d)\exp(j2\phi)\exp(-jk_{x1}d)\exp(j2\phi) \tag{2.16}$$

where ϕ is either ϕ_{TE} or ϕ_{TM}, as appropriate. To achieve self-consistency, $E_s = E_f$, resulting in the *transverse resonance condition*:

$$4\phi - 2k_{x1}d = 2m\pi \tag{2.17}$$

where m, an integer, is called the mode number or rank. For TE waves, (2.14) and (2.17) are used to obtain

$$\tan^{-1}\left(\frac{\gamma_2}{k_{x1}}\right) = \frac{k_{x1}d}{2} + \frac{m\pi}{2} \tag{2.18}$$

or

$$\frac{\gamma_2}{k_{x1}} = \tan\left(\frac{k_{x1}d}{2} + \frac{m\pi}{2}\right)$$

$$= \begin{cases} \tan\left(\dfrac{k_{x1}d}{2}\right) & m = 0,2,4,\ldots \\ -\cot\left(\dfrac{k_{x1}d}{2}\right) & m = 1,3,5,\ldots \end{cases} \tag{2.19}$$

A similar procedure is carried out for the TM case. Collecting results, the following four eigenvalue equations are obtained:

$$\text{even TE:} \quad \frac{\gamma_2}{k_{x1}} = \tan\left(\frac{k_{x1}d}{2}\right) \qquad m = 0,2,4,\ldots \tag{2.20}$$

$$\text{odd TE:} \quad \frac{\gamma_2}{k_{x1}} = -\cot\left(\frac{k_{x1}d}{2}\right) \qquad m = 1,3,5,\ldots \tag{2.21}$$

$$\text{even TM:} \quad \frac{\gamma_2}{k_{x1}} = \frac{n_2^2}{n_1^2}\tan\left(\frac{k_{x1}d}{2}\right) \qquad m = 0,2,4,\ldots \tag{2.22}$$

$$\text{odd TM:} \quad \frac{\gamma_2}{k_{x1}} = -\frac{n_2^2}{n_1^2}\cot\left(\frac{k_{x1}d}{2}\right) \qquad m = 1,3,5,\ldots \tag{2.23}$$

These transcendental equations must be solved graphically or numerically for γ_2 and k_{x1}. β can then be found using (2.10). A graphical technique for their solution is presented in Section 2.4.

2.2. FIELD ANALYSIS OF THE SLAB WAVEGUIDE

In the last section, the eigenvalue equations for the slab waveguide modes were obtained without any direct use of the actual field configurations. In fact, the fields in the slab were *indirectly* taken into account when using the Fresnel equations to obtain the condition for total reflection, in addition to the phase shifts of the reflected rays. In this section, the slab mode fields are found by solving the wave equation. It is then shown how the eigenvalue equations are determined by applying appropriate boundary conditions for the fields at the guide interfaces.

The analysis begins by assuming an electric field solution of the form

$$\mathbf{E}(x, z) = \mathbf{E}_0(x) \exp(-j\beta z) \tag{2.24}$$

where $\mathbf{E}_0(x)$ is to be determined. The wave equation is used, which, for a general z-propagating sinusoidal field, assumes the form given by (1.21):

$$\nabla_t^2 \mathbf{E}_0(x, y) + (k^2 - \beta^2)\mathbf{E}_0(x, y) = 0 \tag{2.25}$$

where ∇_t^2 is the Laplacian taken over x and y only. In cartesian coordinates, (2.25) will separate into individual equations for each component of $\mathbf{E}_0$. In the present case, TM waves will be considered; the z component of $\mathbf{E}_0$ will be found. Since $k^2 - \beta^2 = k_x^2$, and since there is no y variation, (2.25) simplifies to the following expression that is valid inside the slab:

$$\frac{d^2 E_{z1}}{dx^2} + k_{x1}^2 E_{z1} = 0 \tag{2.26}$$

Above and below the slab, $k_x = k_{x2} = -j\gamma_2$ and, as a result,

$$\frac{d^2 E_{z2}}{dx^2} - \gamma_2^2 E_{z2} = 0 \tag{2.27}$$

The solution of (2.26) is:

$$E_{z1}(x, z) = [E_e \sin(k_{x1}x) + E_o \cos(k_{x1}x)] \exp(-j\beta z)$$
$$\frac{-d}{2} \leq x \leq \frac{d}{2} \tag{2.28}$$

where E_e and E_o are the field amplitudes for the cases of even and odd mode number, m, as will be seen later. E_{z1} is seen to be a standing wave in x, characterized by spatial frequency k_{x1} in that direction. The field pattern propagates in the forward z direction with propagation constant β.

The solution of (2.27) is a pair of evanescent waves, existing above and below the slab:

$$E_{z2}(x, z) = \begin{cases} E_c \exp\left[-\gamma_2(x - d/2)\right] \exp\left(-j\beta z\right) & x \geq d/2 \\ \pm E_c \exp\left[+\gamma_2(x + d/2)\right] \exp\left(-j\beta z\right) & x \leq -d/2 \end{cases} \quad (2.29)$$

The field amplitude, E_c, is found in terms of E_e and E_o by applying the boundary condition of continuous tangential electric field at each interface:

$$E_{z1}(x = \pm d/2) = E_{z2}(x = \pm d/2) \quad (2.30)$$

The choice of either sine or cosine for the x dependence of E_{z1} provides two separate cases for the TM modes. Consider first the choice of the sine dependence, which will be called *Case I TM*:

$$E_{z1}^I(x, z) = E_e \sin(k_{x1}x) \exp(-j\beta z) \quad (2.31)$$

Applying (2.30) and using (2.29) results in $E_c = E_e \sin(k_{x1}d/2)$. Outside the slab, the field for Case I TM then assumes the form

$$E_{z2}^I(x, z) = \begin{cases} E_e \sin(k_{x1}d/2) \exp\left[-\gamma_2(x - d/2)\right] \exp(-j\beta z) & x \geq d/2 \\ -E_e \sin(k_{x1}d/2) \exp\left[+\gamma_2(x + d/2)\right] \exp(-j\beta z) & x \leq -d/2 \end{cases}$$

$$(2.32)$$

The minus sign is chosen for the $x \leq -d/2$ field, since $\sin(k_{x1}x)$ changes sign at $x = 0$.

The other possibility is called *Case II TM*:

$$E_{z1}^{II}(x, z) = E_o \cos(k_{x1}x) \exp(-j\beta z) \quad (2.33)$$

Equations (2.29) and (2.30) are used with (2.33) to obtain

$$E_{z2}^{II}(x, z) = \begin{cases} E_o \cos(k_{x1}d/2) \exp\left[-\gamma_2(x - d/2)\right] \exp(-j\beta z) & x \geq d/2 \\ E_o \cos(k_{x1}d/2) \exp\left[+\gamma_2(x + d/2)\right] \exp(-j\beta z) & x \leq -d/2 \end{cases}$$

$$(2.34)$$

A similar procedure can be carried out for TE waves, in which the wave equation is solved for H_z instead of E_z. The steps involved are nearly identical to the above, the major exception being that tangential continuity of H across an interface is used to evaluate constants. Inside the slab, H_z assumes a sine (Case I) or a cosine (Case II) form, in analogy to E_z. The details of the wave equation solution for TE waves are left as an exercise.

Having the z components of $\mathbf{E}$ inside and outside the slab, the transverse components remain to be found. This is accomplished in a straightforward manner using Maxwell's equations. The two curl equations, assuming sourceless media, take the form

$$\nabla \times \mathbf{E} = -j\omega\mu\mathbf{H} \tag{2.35}$$

and

$$\nabla \times \mathbf{H} = j\omega\epsilon\mathbf{E} \tag{2.36}$$

Equations (2.35) and (2.36) can each be separated into three components. For example,

$$(\nabla \times \mathbf{E})_x = \left(\frac{\partial E_z}{\partial y} - \frac{\partial E_y}{\partial z} \right) = -j\omega\mu H_x, \text{ etc.}$$

where $\partial/\partial z = -j\beta$. Continuing this procedure results in a set of six equations representing the individual components of (2.35) and (2.36):

$$\frac{\partial E_z}{\partial y} + j\beta E_y = -j\omega\mu H_x \tag{2.35a}$$

$$\frac{\partial E_z}{\partial x} + j\beta E_x = j\omega\mu H_y \tag{2.35b}$$

$$\frac{\partial E_y}{\partial x} - \frac{\partial E_x}{\partial y} = -j\omega\mu H_z \tag{2.35c}$$

$$\frac{\partial H_z}{\partial y} + j\beta H_y = j\omega\epsilon E_x \tag{2.36a}$$

$$\frac{\partial H_z}{\partial x} + j\beta H_x = -j\omega\epsilon E_y \tag{2.36b}$$

$$\frac{\partial H_y}{\partial x} - \frac{\partial H_x}{\partial y} = j\omega\epsilon E_z \tag{2.36c}$$

Using (2.35a), (2.35b), (2.36a), and (2.36b), it is possible to obtain expressions for individual transverse components in terms of derivatives of z components. For example, E_y can be eliminated between (2.35a) and (2.36b) to obtain

$$H_x = -\frac{1}{k_x^2} \left(j\beta \frac{\partial H_z}{\partial x} - j\omega\epsilon \frac{\partial E_z}{\partial y} \right) \tag{2.37}$$

where it will be recalled that $k_x = (\omega^2\mu\epsilon - \beta^2)^{1/2}$. Proceeding in like manner, equations for the remaining three transverse components are found:

$$H_y = -\frac{1}{k_x^2}\left(j\beta\frac{\partial H_z}{\partial y} + j\omega\epsilon\frac{\partial E_z}{\partial x}\right) \tag{2.38}$$

$$E_x = -\frac{1}{k_x^2}\left(j\beta\frac{\partial E_z}{\partial x} + j\omega\mu\frac{\partial H_z}{\partial y}\right) \tag{2.39}$$

$$E_y = -\frac{1}{k_x^2}\left(j\beta\frac{\partial E_z}{\partial y} - j\omega\mu\frac{\partial H_z}{\partial x}\right) \tag{2.40}$$

As an example, the y component of $\mathbf{H}$ can be found. Using (2.38) (with E_z only) results in

$$
\begin{aligned}
H_y &= -\frac{1}{k_x^2}\left(j\omega\epsilon\frac{\partial E_z}{\partial x}\right) \\
&= \begin{cases} -(1/k_{x1}^2)(j\omega\epsilon_0 n_1^2\,\partial E_{z1}/\partial x) & -d/2 \leq x \leq d/2 \\ (1/\gamma_2^2)(j\omega\epsilon_0 n_2^2\,\partial E_{z2}/\partial x) & x \geq d/2 \text{ and } x \leq -d/2 \end{cases} \tag{2.41}
\end{aligned}
$$

For Case I TM, it is found that

$$H_{y1}^I(x,z) = \frac{-j\omega\epsilon_0 n_1^2}{k_{x1}} E_e \cos(k_{x1}x)\exp(-j\beta z) \tag{2.42}$$

and

$$
H_{y2}^I(x,z) =
\begin{cases}
\dfrac{-j\omega\epsilon_0 n_2^2}{\gamma_2} E_e \sin(k_{x1}d/2)\exp[-\gamma_2(x - d/2)]\exp(-j\beta z) \\
\qquad x \geq d/2 \\
\dfrac{-j\omega\epsilon_0 n_2^2}{\gamma_2} E_e \sin(k_{x1}d/2)\exp[+\gamma_2(x + d/2)]\exp(-j\beta z) \\
\qquad x \leq -d/2
\end{cases}
\tag{2.43}
$$

The requirement of tangential continuity of $\mathbf{H}$ across an interface is now used, which in the present case becomes

$$H_{y1}^I(x = \pm d/2) = H_{y2}^I(x = \pm d/2) \tag{2.44}$$

Substitution of (2.42) and (2.43) into (2.44) leads directly to the eigenvalue equation, (2.22). It is thus seen that Case I corresponds to even values of mode number, m, whereas the Case II solutions will be found to be associated with odd m. Each of Eqs. (2.20) through (2.23) can be derived using procedures similar to that shown above.

2.3. SOLUTIONS OF THE EIGENVALUE EQUATIONS

The solutions of (2.20) through (2.23) are now addressed. It will be found that for each value of m (and for a fixed guide configuration and wavelength), there is a single value of k_{x1} and a single value of β for each of the two mode types (TE and TM). k_{x1} increases as m increases, whereas β decreases. Also, for a given value of m, k_{x1} and β will vary with wavelength, slab thickness, and the refractive indices.

   Equations (2.20) through (2.23) are best solved numerically. On the other hand, the use of a graphical method, while not as accurate, provides a useful illustration of mode behavior. The graphical method presented here uses the x-directed propagation and attenuation constants, k_{x1} and γ_2, that are normalized with respect to the slab thickness, d. By using (2.9) and (2.10), it is found that

$$(k_{x1}d)^2 + (\gamma_2 d)^2 = k_0^2 d^2(n_1^2 - n_2^2) \tag{2.45}$$

When plotted with coordinate axes, $k_{x1}d$ and $\gamma_2 d$, (2.45) is the equation of a circle of radius

$$R = k_0 d(n_1^2 - n_2^2)^{1/2} \tag{2.46}$$

Solving (2.45) for $\gamma_2 d$ and using (2.20) and (2.21) lead to

$$\gamma_2 d = d[k_0^2(n_1^2 - n_2^2) - k_{x1}^2]^{1/2} = \begin{cases} k_{x1}d \tan(k_{x1}d/2) & m = 0, 2, 4, \ldots \\ -k_{x1}d \cot(k_{x1}d/2) & m = 1, 3, 5, \ldots \end{cases}$$

$$\tag{2.47}$$

for TE modes. The above is plotted in Fig. 2.4. Note that the tangent and cotangent curves get steeper with branch number, since each is multiplied by $k_{x1}d$. The intersections of the circle with the branches determine solutions (values of k_{x1} and γ_2) of (2.47).

   Several facts can be observed from Fig. 2.4.

1. From (2.46) it is seen that an increase in frequency, index difference, or slab thickness will increase the value of R, leading to additional solutions and thus the possibility of exciting more modes.

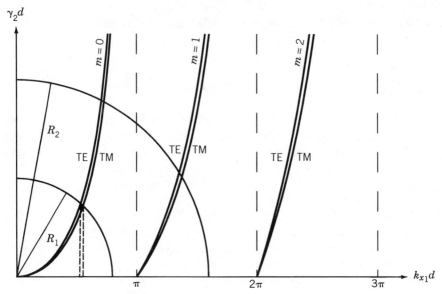

Figure 2.4. Plot of a graphical solution for TE and TM modes in a symmetric slab waveguide, showing the single-mode (R_1) and multimode (R_2) cases. Note that the TE and TM modes that are associated with a given value of m are nearly degenerate, meaning that their k_{x1} and β values are nearly equal.

2. Cutoff for a mode of mode number m occurs when $R = m\pi$. At this value:

   a. The cutoff frequency can be determined from the above to be

$$f_c = \frac{\omega_c}{2\pi} = \frac{mc}{2d(n_1^2 - n_2^2)^{1/2}} \qquad (2.48)$$

   where c is the velocity of light. At frequencies less than f_c, the mode will not propagate.

   b. The value of γ_2 is zero, indicating no confinement for the mode.

   c. The ray angle, θ_1, can be shown to be equal to the critical angle, θ_c.

3. There is no cutoff for an $m = 0$ mode.

4. At a given frequency, k_{x1} is larger and θ_1 is less for higher order modes (recall that $k_{x1} = n_1 k_0 \cos \theta_1$).

5. At a given frequency, γ_2 is greater for lower order modes. As a result, these are more tightly confined to the slab.

Figure 2.4 depicts a graphical solution for TE and TM modes. Including the TM modes introduces an additional set of tangent and cotangent curves that are slightly offset from the TE curves as shown; the offset arises from the added factor of n_2^2/n_1^2 from (2.23) and (2.24). Each value of m therefore corresponds to a pair of modes, TE_m and TM_m, which will have propagation constants that differ slightly. It is thus impossible to obtain a truly single-mode symmetric slab guide by lowering the frequency to obtain $m = 0$ only.

A final point concerns the behavior of a given mode with frequency. Referring again to Fig. 2.4, note that two circles are shown, where R_2 could be said to differ from R_1 by a simple increase in frequency. R will increase linearly with frequency, whereas it is seen that $k_{x1}d$ (two values of which are shown here for the $m = 0$ mode pair) will increase at a slower rate. Noting that $k_{x1} = n_1 k_0 \cos \theta_1$, we see that a linear increase in frequency (k_0) must therefore be accompanied by a *decrease* in $\cos \theta_1$ in order to achieve this behavior. Consequently, for a given mode, θ_1 will *increase* with frequency. Since $\beta = n_1 k_0 \sin \theta_1$, it is found that the phase velocity—$v_p = \omega/\beta = c/(n_1 \sin \theta_1)$—will therefore decrease as frequency increases.

2.4. POWER TRANSMISSION AND CONFINEMENT

An important attribute of any waveguide is the amount of power transmitted through it, and the manner in which the power is distributed within the structure. In dielectric guides, the existence of fields outside the slab region can result in outside objects affecting power transmission. These effects are minimized if the mode field amplitudes at the object location are as small as possible, a condition achieved if the outer fields exhibit rapid exponential decay in x. On the other hand, in device applications involving slab waveguides, the evanescent power in the cladding region is typically used for input and output coupling, as well as to enable the guided power to interact with other components that can be conveniently placed on top of the slab structure. In the case of optical fibers, knowledge of the power distribution is of critical importance for minimizing losses and for optimizing input and output coupling efficiencies. These issues will be addressed in later chapters.

Power computation begins with the time-average power density as expressed in (1.24):

$$\langle \mathbf{S} \rangle = \tfrac{1}{2} \operatorname{Re} \{ \mathbf{E} \times \mathbf{H}^* \} \quad \text{W/m}^2 \tag{2.49}$$

where $\mathbf{S}$ is the Poynting vector, and where the electric and magnetic fields are in phasor form. The average power that is guided by the slab waveguide will be the integral of the z component of $\mathbf{S}$ over the guide transverse cross section. The general expression for the power will thus be

$$P = \int_{-\infty}^{\infty} \int_{-\infty}^{\infty} \langle \mathbf{S} \rangle \cdot \hat{\mathbf{a}}_z \, dx \, dy \quad \text{W} \tag{2.50}$$

The above would give infinite power since the guide and fields are assumed of infinite extent in the y direction. Considering a unit width in y, and noting that only the transverse components of $\mathbf{E}$ and $\mathbf{H}$ will take part in the z-directed power flow, the time-average power per unit width will be

$$P = \frac{1}{2} \int_{-\infty}^{\infty} \mathrm{Re}\,\{E_x H_y^* - E_y H_x^*\}\, dx \quad \text{W/m} \tag{2.51}$$

In practice, a beam of light having finite diameter is typically coupled into the guide, leading to a mode of finite y dimension, with y dependence (such as gaussian) determined by that of the original input beam. The complete calculation expressed by (2.50) would then be used to find the total power.

It was shown in the last section that the shape of a given mode will change as frequency is varied, in that k_{x1} and γ_2 change with the circle radius, R, on the graphical solution. Increasing γ_2 leads to more rapid exponential decay of the mode fields away from the slab. As a result, the *power confinement* within the guide increases. The latter quantity can be expressed as the ratio of the power that propagates within the slab region to the total power:

$$\frac{P_{\text{slab}}}{P_{\text{slab}} + P_{\text{clad}}} = \frac{\displaystyle\int_{-d/2}^{d/2} \mathrm{Re}\{E_{x1} H_{y1}^* - E_{y1} H_{x1}^*\}\, dx}{\displaystyle\int_{-\infty}^{\infty} \mathrm{Re}\,\{E_x H_y^* - E_y H_x^*\}\, dx} \tag{2.52}$$

In the symmetric slab,

$$P_{\text{clad}} = (2)\frac{1}{2} \int_{d/2}^{\infty} \mathrm{Re}\,\{E_{x2} H_{y2}^* - E_{y2} H_{x2}^*\}\, dx \tag{2.53}$$

The factor of 2 in the above accounts for the region $-\infty < x < -d/2$, which by symmetry contains the same amount of power as that in the region $x > d/2$. The confinement as expressed in (2.52) reaches its maximum value of unity at infinite graph radius, since γ_2 approaches infinity, and P_{clad} thus approaches zero. From similar reasoning, (2.52) is zero at cutoff, since $\gamma_2 = 0$; in this case the fields are of uniform magnitude over the entire range of x. The power within the slab is therefore negligible compared to P_{clad}.

Another property of power transmission has to do with its distribution within the various modes. The modes of any waveguide form an *orthogonal set*. In the case of the slab waveguide, this property is stated as

$$\int_{-\infty}^{\infty} \text{Re} \{\hat{\mathbf{E}}_q \times \hat{\mathbf{H}}_p^*\} \cdot \hat{\mathbf{a}}_z \, dx = \delta_{qp} \qquad (2.54)$$

where q and p are values of the mode number, m. δ_{qp} is the Kronecker delta, having a value of one when $q = p$, and zero otherwise. Equation (2.54) states that the flow of power is restricted to the individual modes; ultimately the total power will be the sum of the power contributions in each mode. $\hat{\mathbf{E}}_q$ and $\hat{\mathbf{H}}_p$ are the electric and magnetic fields of modes q and p, which contain the transverse dependences only (the $\exp(-j\beta z)$ terms are omitted). The fields are appropriately normalized to yield a value of one for (2.54) when $q = p$.

2.5. LEAKY WAVES

In addition to the purely guided modes, other field configurations exist in dielectric guides that form propagating waves that exhibit what could be thought of as temporary confinement. Leaky waves can be described through the reflecting ray picture as waves that satisfy the transverse resonance condition, but that partially transmit through the interfaces at each reflection. Power in the slab is thus lost as the wave propagates. A consequence of this behavior is that leaky waves are not true modes of the structure. The power loss from the slab in fact arises from the spatially transient redistribution of the wave power from the slab to the surrounding regions. Thus the shape of the total field over the guide cross section is continually changing with z and does not meet the basic definition of a mode, which specifies that the field distribution in space will be constant with z, allowing only for a uniform overall loss. Leaky wave field distributions, while changing with z, are constant in time. An excellent discussion of the physical characteristics of leaky waves is found in ref. 3.

In the symmetric slab guide, leaky waves can exist if either of two structural conditions are met: (1) the slab index is less than the cladding index, so that total reflection from within is not possible; or (2) the slab index is greater than the cladding index as with purely guided modes, but the ray incidence angle is less than the critical angle. The two cases lead to two eigenvalue equations that differ slightly from one another. But, in contrast to the purely guided mode case, both can be solved analytically. The two cases are described by the generalized diagram of Fig. 2.3. Since the waves are leaky, the x component of $\mathbf{k}_2$ will be real, whereas it was imaginary for guided modes. This means that values of β for leaky waves are restricted to the range $\beta < n_2 k_0$. Since the incident ray from within the slab can be TE or TM polarized as before, the appropriate Fresnel equation, (2.5) or (2.6), must be used to determine the reflected field amplitude and phase shift.

Consider a TE polarized field within a slab guide in which $n_1 < n_2$. From Snell's law, it is found that $\theta_2 < \theta_1$, and therefore $k_{x1} < k_{x2}$. From (2.5), it is seen that Γ_{TE} is thus real and negative, which means that the incident wave will experience a phase shift of π radians on reflection—independent of the incidence angle. Following the same reasoning used in Section 2.2, an equation analogous to (2.16) can be written, which states the transverse resonance requirement for a TE leaky wave in a slab waveguide of thickness d.

$$\exp(-jk_{x1}d)\exp(j\pi)\exp(-jk_{x1}d)\exp(j\pi) = \exp(-j2m\pi) \qquad (2.55)$$

The above condition states that the phase shift over one transverse round trip must be an integer multiple of 2π. The field amplitude, however, will not be maintained due to reflective losses. Equating the arguments of the above exponential functions results in the eigenvalue equation:

$$k_{x1}d = (m + 1)\pi \qquad n_1 < n_2 \qquad (2.56)$$

If the guide is constructed with $n_1 > n_2$, and if $\theta_1 < \theta_c$, then $\theta_1 < \theta_2$, and $k_{x1} > k_{x2}$. Under these conditions, (2.5) indicates that Γ_{TE} will be real and positive, leading to zero phase shift on reflection. Applying the transverse resonance requirement to this case leads to the same eigenvalue equation, but without the $m = 0$ mode possibility:

$$k_{x1}d = m\pi \qquad n_1 > n_2 \qquad (2.57)$$

The analysis for TM waves proceeds similarly, except (2.6) is used to determine the phase shift. It is left as an excercise (Problem 2.8) to show that eigenvalue equations for TM leaky waves result that are identical to (2.56) and (2.57) for the two structural cases. TE and TM leaky waves having the same m value are thus degenerate.

The significance of m is the same for the leaky and guided wave cases. In the $n_1 > n_2$ structure, a wave associated with a value of m that propagates with angle greater than the critical angle forms a guided mode; if the ray angle is less than θ_c, then the mode propagates as a leaky wave. At the critical angle, the eigenvalue equations for the guided and leaky cases become identical. Leaky waves can thus be interpreted as the manifestations of guided modes below cutoff. The absence of the $m = 0$ leaky wave in the $n_1 > n_2$ structure results since the $m = 0$ guided mode has no cutoff; it thus has no leaky counterpart.

An important characteristic of a leaky structure is the amount of power that is lost by transmission though the dielectric interfaces as the slab wave propagates. This loss can be calculated over a unit distance in z and then compared to the power in the slab at the input. The power at the output of a unit length of guide, P_o, can be expressed in terms of the input power and the power loss coefficient, 2α, through

$$P_o = P_i \exp(-2\alpha) = P_i - P_l \qquad (2.58)$$

The power lost per unit length is therefore:

$$P_l = P_i[1 - \exp(-2\alpha)] \approx 2\alpha P_i \qquad (2.59)$$

where the approximation is valid for $2\alpha \ll 1$.

The power densities in the plane waves are found through their Poynting vectors $\mathbf{S}$, which lie in the direction of $\mathbf{k}_u$ and $\mathbf{k}_d$, as shown in Fig. 2.1. Consider upward and downward plane waves with TE polarization that propagate in a section of the slab having unit width in y and unit length in z. Since the slab guide is symmetric, the waves will carry equal power densities of magnitude S in each. The incident power per unit area on each dielectric interface at $\pm d/2$ will be the product of the x-directed power density in either one of the waves and the area corresponding to unit length in z and unit width in y:

$$P_d = 2\mathbf{S} \cdot \hat{\mathbf{a}}_x = 2S \cos \theta_1 = 2S \frac{k_{x1}}{n_1 k_0} \qquad (2.60)$$

where the factor of 2 arises from the existence of two waves and two interfaces. The z-directed power flow of the two waves in the slab will be

$$P_i = 2d\mathbf{S} \cdot \hat{\mathbf{a}}_z = 2dS \sin \theta_1 = 2dS \frac{\beta}{n_1 k_0} \qquad (2.61)$$

This result, obtained by the simple product of the z-directed power densities of the plane waves with the slab cross-sectional area, is in fact consistent with the method outlined in Section 2.4. There, the net mode fields are used and an integral is taken over the guide cross section. The consistency is understood by recalling that the mode fields are in fact formed by the interference of the two plane waves; thus the power flow can be thought of as appearing in either form, and the two computation methods must yield the same result.

The power lost by transmission through the interfaces is now

$$P_l = P_d(1 - |\Gamma_{TE}|^2) \qquad (2.62)$$

where Γ_{TE} is given by (2.5). The power loss coefficient becomes

$$2\alpha \approx \frac{P_l}{P_i} = \frac{k_{x1}}{\beta d}(1 - |\Gamma_{TE}|^2) \qquad (2.63)$$

Substituting (2.5) results in

$$2\alpha \approx \frac{4k_{x1}^2 k_{x2}}{\beta d(k_{x1} + k_{x2})^2} \qquad (2.64)$$

where, again, the above approximation is valid as long as $2\alpha \ll 1$. This is accomplished by selecting a unit distance in z for the calculation that is sufficiently short. The same procedure is carried out for TM waves, using (2.6), that yields a similar result (Problem 2.10).

2.6. RADIATION MODES

Consider a point source located within the slab region, the radiation from which will initially propagate with uniform amplitude in all directions. Part of the emitted power will propagate in directions appropriate for coupling into the leaky or guided modes. But it is clear that additional field solutions for the structure are needed to account for the radiated waves that propagate at angles other than those associated with the modes. The added solutions are known as the *radiation modes*. In contrast to the leaky and guided waves, radiation modes have no transverse resonance requirement and always appear in continua at a given frequency—in other words, in groups within which β varies continuously. They are of two types: (1) waves that propagate out of the slab, having angles less than the critical angle; and (2) fields that exhibit exponential decay (but no propagation) in the z direction.

The radiation modes, together with the guided modes, form a *complete set*—that is, any field distribution in the guide structure that satisfies Maxwell's equations can be expressed as an expansion of guided and radiation modes. Leaky waves are not a part of this set since, as demonstrated above, they are transient in z. In contrast to the leaky and guided modes, to which are associated discrete ray angles and β values, it is not possible to excite a single radiation mode from within the slab. This would require a plane wave source inside the slab of infinite extent. Instead, radiation modes are always excited in groups that exhibit continuous variation in β over a certain range; the range of β is determined by the source angular radiation characteristics.

An elegant discussion of radiation modes in asymmetric slab waveguides is presented in ref. 1. The heart of the description lies in recognizing that the slab guide can exhibit metallic waveguide properties in addition to those already demonstrated. To see this, consider the placement of perfectly conducting surfaces on the top and bottom of the slab. The structure thus becomes a parallel-plate waveguide, in which the free-space cutoff wavelength for mode m is $\lambda_c = (2n_1 d)/(m + 1)$. This can easily be shown by writing down the transverse resonance condition with a π phase shift at each reflection. In terms of λ_c, the propagation constant will be

$$\beta = n_1 k_0 \sin\theta_1 = n_1 k_0 \sqrt{1 - (\lambda/\lambda_c)^2} \qquad (2.65)$$

where the allowed range of θ_1 is between 0 and $\pi/2$, and where cutoff occurs when $\theta_1 = 0$. At cutoff for mode m, there will be exactly $m+1$ half-wavelengths (measured vertically in the slab material) between the plates. Note that while β is real for wavelength values that are less than λ_c, the wave is evanescent in z (exhibits exponential decay with no propagation) when $\lambda > \lambda_c$, since β is imaginary over the latter range.

Next, suppose the plates are allowed to separate from the slab and recede toward positive and negative infinity in x. As this happens, three conditions arise: (1) the dielectric slab waveguide modes return, essentially unperturbed, since the plates will have little effect on the evanescent slab guide fields if they are positioned at far enough distances; (2) the parallel-plate modes continue to exist, but with the three media $(n_2 : n_1 : n_2)$ between plates; and (3) the cutoff wavelength separation between adjacent parallel-plate modes decreases as the plate separation increases and reaches zero as the plates recede to infinity. In the latter case, the parallel-plate modes will form a continuum of allowed β values. These modes, above cutoff, are identified as the first type of radiation modes in the slab waveguide described above, since they propagate through the slab region at angles that are always less than the critical angle. Radiation modes of the second type, evanescent in z, are interpreted as parallel-plate modes below cutoff. These can be used, for example, to more accurately model the fields in the vicinity of a waveguide imperfection. Such modes will always exist, since a sufficiently high value of m can be found to place the mode associated with it below cutoff at a given wavelength, regardless of the plate separation.

The grouping of the various wave types in the symmetric slab guide is summarized in the ω–β diagram shown in Fig. 2.5. The key boundaries are lines of constant phase velocity, having slopes $\omega/\beta = c/n_1$ and c/n_2. Guided modes, for example, thus lie between these boundaries, as specified by (2.12). The for-

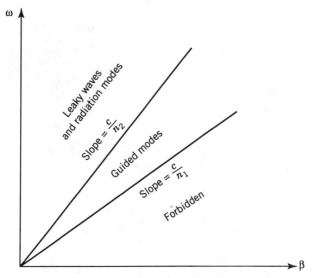

Figure 2.5. Propagation constant grouping for the wave types in a symmetric slab waveguide.

bidden zone would contain waves having phase velocities slower than c/n_1, a condition not possible for any wave in the slab guide.

2.7. WAVE PROPAGATION IN CURVED SLAB WAVEGUIDES: RADIATION LOSS

In the previous sections, the waveguiding structure has been assumed straight. One consequence of this is that modes exist that are fully guided and thus experience no loss from power radiating from the structure. A different situation arises when the guide is curved, accomplished either by design or by bending an otherwise straight guide. In this case, it is no longer possible to achieve complete confinement of any mode. Some fraction of the power will be radiated away to an extent that depends on the usual waveguide structural parameters, in addition to the wavelength and the bend radius, r_b.

Figure 2.6 illustrates the geometry of wave propagation in a curved slab waveguide. Recall that in the straight guide, ray incidence on the slab–cladding interface at angles greater than the critical angle results in total reflection. In the

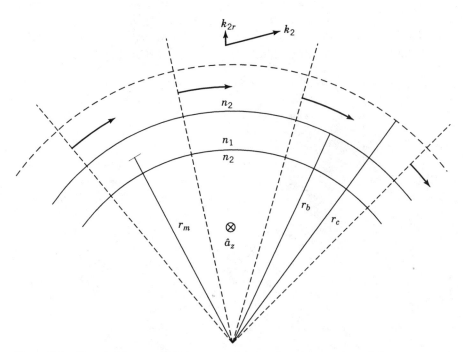

Figure 2.6. Curved slab waveguide geometry, showing the existence of a radial component of $\mathbf{k}_2$ (k_{2r}) at distances beyond the critical radius, r_c.

cladding region, a surface wave exists that propagates at phase velocity ω/β in the z direction. In the curved structure, the surface wave will also exist when $\theta_1 > \theta_c$. The difference is that all constant phase surfaces in the outside wave will lie along radial lines, normal to the guided mode propagation direction as required. To maintain this, the phase velocities of the surfaces must increase linearly with increasing radius. The propagation constant, β, will thus be in proportion to $1/r$ and can be written explicitly as

$$\beta(r) = \frac{r_b}{r}\,\beta(r_b) \tag{2.66}$$

At a value of r known as the *critical radius*, the wave velocity becomes equal to that of light in the cladding material. It is readily shown (Problem 2.11) that for incident angles greater than the critical angle, the critical radius, r_c, relates to the bend radius, r_b, through

$$r_b \sin \theta_1 = r_c \sin \theta_c \tag{2.67}$$

where $\sin \theta_c = n_2/n_1$ as before. For incident angles less than or equal to θ_c, $r_c = r_b$.

At radii larger than r_c, the wave velocity in the guide direction exceeds that of light. As a result, the propagation vector in the outer cladding region, $\mathbf{k}_2$, establishes a radial component at locations beyond the critical radius. This effect allows phase velocity along the guide to exceed that of light but limits the velocity along the direction of power flow (along $\mathbf{k}_2$) to that of light in the cladding as is required. The power is thus no longer confined to the guide and radiates away. Since β is proportional to $1/r$, it approaches zero at infinite radius, and thus k_2 at infinity should exhibit only a radial component. One would expect to observe this behavior in the field solutions of the wave equation. Increasing θ_1 beyond the critical angle at a curved interface thus does not yield total reflection; instead, it results in an increase in the critical radius, at which the refracted wave finally radiates. This phenomenon of an incident wave emerging to radiate at a different radial location is known as *optical tunneling*.

A number of insights are gained by considering the reflection and transmission of rays at the outer curved interface. Incidence on curved surfaces was analyzed in detail by Snyder and Love [4], the primary results of which lead to a generalization of the Fresnel equations for reflection and transmission. The latter were derived in Chapter 1 for the special case of ray incidence on plane interfaces.

The Snyder and Love model treats the curved boundary as a perturbation of a plane interface by assuming that the radius of curvature is much larger than a wavelength. In so doing, all wavefronts can be considered essentially plane within local regions in the vicinities of the dielectric interface and the critical radius. Cartesian coordinates can be used, which simplifies the analysis.

Referring to Fig. 2.7, the origin is positioned at the boundary, with increases in radius beyond r_b measured by x, and with the critical radius located at $x = x_c$. z is tangent to the boundary. The propagation constant can be expressed in terms of x by modifying (2.66) to read

$$\beta(x) \approx \beta(0)[1 - (2x/r_b)]^{1/2} \tag{2.68}$$

where it is assumed that $x \ll r_b$. Since $k_2 = n_2 k_0$ is a constant, the x-directed propagation constant in medium 2 is readily found through

$$k_{x2}^2(x) = k_2^2 - \beta^2(x) \tag{2.69}$$

With the fields assumed to vary only with x, the wave equation reduces to the form exemplified in (2.26) and (2.27). In medium 2 the wave equation reads as follows for the TE case:

$$\frac{d^2 E_{y2}(x, z)}{dx^2} + k_{x2}(x) E_{y2}(x, z) = 0 \tag{2.70}$$

where the assumed form of the electric field is

$$E_{y2}(x, z) = E_{y2}(x) \exp[-j\beta(x)z] \tag{2.71}$$

Substituting (2.71) into (2.70) and invoking a change of variable [4] result in Airy's equation:

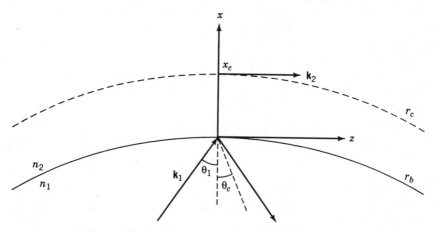

Figure 2.7.   Ray geometry at a curved dielectric interface.

$$\frac{d^2 E_{y2}(\xi)}{d\xi^2} - \xi(x)E_{y2}(\xi) = 0 \tag{2.72}$$

where

$$\xi(x) = -\left(\frac{2\beta^2(0)}{r_b}\right)^{-2/3}\left(k_{x2}^2(0) + 2\frac{\beta^2(0)x}{r_b}\right) \tag{2.73}$$

Equation (2.72) is solved to yield $E_{y2}(x)$. The net field is then constructed using (2.71):

$$E_{y2}(x,z) = E_{y0}\text{Ai}\left[\xi(x)\exp(j2\pi/3)\right]\exp\left[-j\beta(x)z\right] \tag{2.74}$$

where $\text{Ai}\left[\xi \exp(j2\pi/3)\right]$ is the linear combination of Airy functions defined through

$$\text{Ai}\left[\xi(x)\exp(j2\pi/3)\right] \equiv \tfrac{1}{2}\exp(j\pi/3)[\text{Ai}(\xi) - j\text{Bi}(\xi)] \tag{2.75}$$

The Airy functions, $\text{Ai}(\xi)$ and $\text{Bi}(\xi)$, are plotted in Fig. 2.8.

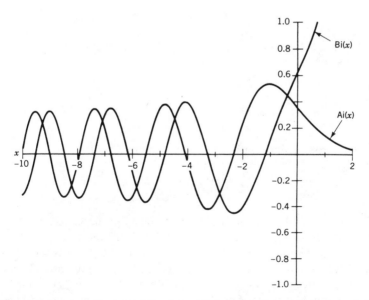

Figure 2.8.   Airy functions. From C. M. Bender and S. A. Orszag, *Advanced Mathematical Methods for Scientists and Engineers* (McGraw-Hill, New York, 1978). Reprinted with permission of McGraw-Hill.

Detailed discussions of these functions can be found elsewhere [5,6]. Of immediate importance in this study is that they exhibit either oscillatory or exponential-like behavior depending on whether ξ is less than or greater than zero. If the functions are oscillatory, E_{y2} will propagate with an x component of $\mathbf{k}_2$. If they are exponential, E_{y2} will exhibit exponential variation with x in the manner of a confined mode.

At the critical radius, $\beta(x_c) = n_2 k_0$. By using (2.68), it can be shown that x_c is given by

$$x_c = r_b \frac{\beta^2(0) - n_2^2 k_0^2}{2\beta^2(0)} \tag{2.76}$$

Then, by using $k_{x2}^2(0) = n_2^2 k_0^2 - \beta^2(0)$, it is found from (2.73) that $\xi(x_c) = 0$. The critical radius thus corresponds to the transition point between exponential $(x < x_c)$ and oscillatory $(x > x_c)$ behavior of the electric and magnetic field solutions with x.

The fraction of the incident power transmitted through the interface is found by deriving reflection and transmission coefficients, as was done for plane interfaces (Chapter 1). Requiring continuity of the tangential electric and magnetic field components at the curved boundary $(x = 0)$ leads to modified versions of the Fresnel equations, (2.5) and (2.6). Resulting is the power transmission coefficient for curved interfaces [4]:

$$T = C[1 - |\Gamma|^2] \tag{2.77}$$

where Γ is either Γ_{TE} (Eq. (2.5)) or Γ_{TM} (Eq. (2.6)) as appropriate. The function C accounts for the curvature and is given by

$$C = \frac{|Ai[\xi(0) \exp(j2\pi/3)]|^{-2}}{4\pi|\xi(0)|^{1/2}} \tag{2.78}$$

A useful approximation to (2.78) is valid when $\pm\xi \gg 1$.

$$C \approx \exp(-4\xi(0)^{3/2}/3) \tag{2.79}$$

The $\xi(0) \approx 0$ condition applies to incident angles in the vicinity of the critical angle, at which the exact form of C given by (2.78) must be used.

By noting the variation of ξ with r_b as given by (2.73), it is seen that a substantial increase in transmissivity results as bend radius decreases. This has dramatic consequences regarding power loss as a result of propagation through bent guides. The important case of mode-dependent loss in bent fibers will be considered in Chapter 4. The curved interface results also have a direct bearing on the mode types that can exist in round fibers, as will be seen in the next chapter.

PROBLEMS

2.1. Derive the eigenvalue equations, (2.20) and (2.21), for odd and even TE modes in a symmetric slab waveguide. Begin by solving the wave equation for the z component of magnetic field; then obtain (2.20) and (2.21) by evaluating the transverse field components and using the property of continuity of tangential electric field across either dielectric interface.

2.2. A symmetric guide has slab index $n_1 = 1.50$ and surrounding index $n_2 = 1.46$. The slab thickness is $5.0\,\mu m$.

   a) Determine the maximum free-space wavelength at which the $m = 1$ mode will propagate.

   b) Find the number of modes that will propagate when $\lambda = 1.06\,\mu m$.

2.3. Using the graphical solution for the slab waveguide, show that the ray angle, θ_1, at cutoff is equal to the critical angle of total reflection at the slab–cover interface.

2.4. Two slab guides are constructed of identical materials, but the slab thickness of one guide is exactly twice that of the other. Considering the graphical solution for each guide, determine the following:

   a) Which guide has the higher cutoff frequency for the $m = 1$ mode.

   b) Which guide propagates the $m = 0$ mode at the larger ray angle at frequency ω.

   c) Which guide carries the higher percentage of power in the $m = 0$ mode within the core region.

2.5. Two slab guides are constructed identically, except the slab index of one is twice that of the other. Under this condition, reanswer the questions of Problem 2.4.

2.6. A symmetric slab guide propagates a wave for which the z-directed phase velocity of the lowest order mode is one-half the speed of light in vacuum. Assuming a TE mode, find an expression for n_2, in terms of n_1, d, and λ. If $n_1 = 2.1$, determine the allowable range of values of the ratio d/λ, which produce meaningful values of n_2. *Hint:* Start by calculating the required value of the ray angle, θ_1, from which β and k_{x1} can be obtained. Then use the appropriate eigenvalue equation.

2.7. A nonsymmetric slab waveguide is constructed having film index n_1, cover index n_3 ($x > d/2$), and substrate index n_2 ($x < -d/2$), where $n_1 > n_2 > n_3$.

   a) Using transverse resonance arguments, determine the eigenvalue equation for TE mode propagation in this guide.

   b) Determine the minimum ray angle that a guided mode could have in this structure.

2.8. One way to achieve true single-mode operation in a slab guide is to fabricate the guiding layer on a perfectly conducting substrate, as shown in Fig. P2.1. Only two mode types (instead of the usual four) are allowed to propagate in this structure, because of the requirement of zero tangential electric field at the conductor surface.

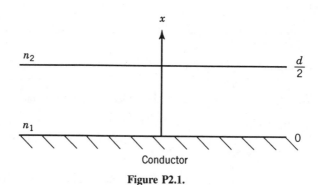

Figure P2.1.

a) Determine which two types of mode are allowed.

b) Verify your above result using transverse resonance arguments.

c) If $d/2$ is equal to the free-space wavelength and if $n_2 = 1.50$, find the maximum value of n_1 that will enable single-mode operation.

d) Find the range of n_1 within which only two modes will propagate.

2.9. Assuming $n_2 < n_1$, determine an expression for the number of leaky modes that will propagate in a symmetric guide. Express your result in terms of the guide thickness, d, wavelength, λ, and indices, n_1 and n_2. Note that leaky modes will occupy the range of ray angles, θ_1, between zero and the critical angle. Repeat the above for the case in which $n_2 > n_1$.

2.10. Derive the power loss formula (analogous to (2.64)) for TM leaky waves in a symmetric guide.

2.11. Derive Eq. (2.67) by finding the critical radius for given values of θ_1, θ_c, and r_b. *Hint:* Write β as a function of radius and use the requirement that tangential phase velocity be equal to that in the cladding material at the critical radius.

REFERENCES

1. D. Marcuse, *Theory of Dielectric Optical Waveguides*, 2nd ed. Academic Press, New York, 1990. Chapter 1.

2. H. Kogelnik, "Theory of Dielectric Waveguides," in *Integrated Optics*, 2nd ed. T. Tamir, ed. Springer-Verlag, New York, 1979.

3. A. A. Oliner, "Historical Perspectives on Microwave Field Theory," IEEE Transactions on Microwave Theory and Techniques, vol. MTT-32, pp. 1022–1045, 1984.

4. A. W. Snyder and J. D. Love, "Reflection at a Curved Dielectric Interface: Electromagnetic Tunneling," *IEEE Transactions on Microwave Theory and Techniques*, vol. MTT-23, pp. 134–141, 1975.

5. C. M. Bender and S. A. Orszag, *Advanced Mathematical Methods for Scientists and Engineers*. McGraw-Hill, New York, 1978.

6. M. Abramowitz and I. Stegun, *Handbook of Mathematical Functions*. Dover, New York, 1965.

3

Optical Fibers with Single-Step Index Profiles

The single-step index profile is the simplest fiber design that employs homogeneous dielectrics, and it thus provides a logical starting point in the study of fibers of more complicated design. This basic profile was seen in most of the early manufactured fibers and is still found in many that are currently made. Although more sophisticated profile designs have largely replaced the single-step, the analysis of these is often simplified by referring to an "equivalent" step index fiber, whose light-guiding properties are similar. This association is possible, since the single-step profile analysis yields qualitative mode characteristics and mode designations that apply to fibers of any profile in which the refractive index maximizes on the fiber axis.

The single-step index fiber (hereafter referred to as step index) consists of two concentric homogeneous dielectrics, as shown in Fig. 3.1. The inner dielectric, or core, fills the region $r < a$ and is of refractive index n_1. The outer material $(r > a)$, known as the cladding, is of index n_2, where $n_2 < n_1$. The cladding is considered, for purposes of analysis, to be of infinite outer radius. Key struc-

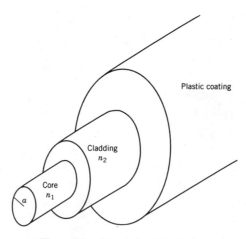

Figure 3.1. Step index fiber design.

tural parameters include the core radius, a, and the normalized index difference (usually expressed in percent):

$$\Delta = \frac{n_1^2 - n_2^2}{2n_1^2} \approx \frac{n_1 - n_2}{n_1} = \frac{\delta n}{n_1} \tag{3.1}$$

The approximation in (3.1) arises from assuming that n_1 is very close in value to n_2, which is almost always true. Another important quantity is the *numerical aperture*, *N.A.*, defined as the sine of the maximum angle (measured outside the fiber with respect to the z axis) of an incident light ray that becomes totally confined within the fiber. By using simple geometrical arguments (left as an exercise) it can be shown that this maximum angle is achieved when the interior ray propagates at the critical angle, measured at the core–cladding interface. By using this fact, along with the relation $\sin \theta_c = n_2/n_1$, it is found that

$$N.A. = \sqrt{n_1^2 - n_2^2} = n_1\sqrt{2\Delta} \tag{3.2}$$

Single-mode and multimode step index fibers form two classifications of practical concern. Standards on dimensions have been set by various committees. Among these are the Electronic Industries Association (EIA), the International Electrotechnical Commission (IEC), and the Comité Consultatif International Télégraphique et Téléphonique (CCITT). All have specified a cladding diameter of 125 μm as the production standard for most single-mode and multimode fibers. Core diameters vary between 50 and 100 μm for multimode, and between 4 and 10 μm for single-mode fibers. Values of Δ are typically about 0.2% for single mode and about 1% for multimode.

As mentioned, the step index profile has in many instances been replaced by more complicated profiles for special purposes. In particular, multimode fibers with graded index profiles exhibit significantly increased bandwidth over the step index case. Alternative profiles for single-mode fibers have become important for minimizing dispersion over specified wavelength ranges and for control of mode polarization. Some of these special designs will be considered in Chapter 6. Their mode field characteristics and frequency behavior are qualitatively similar to those found in the step index profile, and the mode designation system is the same.

3.1. RAY PROPAGATION IN THE STEP INDEX FIBER

In studying the slab waveguide, it was found that ray propagation studies enabled one to derive the eigenvalue equations through transverse resonance, in addition to being able to obtain a physical picture of the mode field solutions. In the slab and fiber, the precise mode fields can in fact be constructed through an appropriate superposition of plane waves (characterized by rays). This procedure becomes complicated, however, when dealing with core sizes on the order of a wavelength. In the present treatment, a simple ray picture will be used to obtain qualitative features of the modes that will be recognized and used later when solving the wave equation.

Cylindrical coordinates are used, with the fiber axis coincident with the z axis. Radius r, the azimuthal angle ϕ, and z are the three coordinates; their respective directions are specified by unit vectors, $\hat{\mathbf{a}}_r$, $\hat{\mathbf{a}}_\phi$, and $\hat{\mathbf{a}}_z$. As in the slab guide, a guided mode in a fiber can be constructed from a group of rays that totally reflect from the interface at $r = a$. Ray propagation in a fiber is complicated by the possibility of a path component in the ϕ direction, thus giving rise to a *skew* ray. Such a ray exhibits a spiral-like path down the core, never crossing the z axis. A *meridional* ray is one that has no ϕ component; it passes through the z axis and is thus in direct analogy to a slab guide ray.

Figure 3.2 shows a skew ray at radius r in the core, along with its components in the three coordinate directions. The ray has propagation constant magnitude $k_1 = n_1 k_0$. This value is related to the component magnitudes by

$$\beta_r^2 + \beta_\phi^2 + \beta^2 = n_1^2 k_0^2 \tag{3.3}$$

The net transverse propagation constant is written as $\beta_t^2 = \beta_r^2 + \beta_\phi^2$. The ϕ component, providing a measure of the skewness, is restricted by the requirement that when multiplied by the core circumference at radius r, it yields a net phase

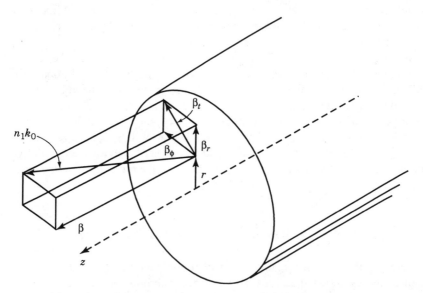

Figure 3.2.   Skew ray decomposition in the core of a step index fiber.

shift of $2q\pi$, where q is an integer. Its value will therefore be $\beta_\phi = q/r$. The z component, β, is the propagation constant of the fiber mode that is associated with the ray.

In analogy to the slab waveguide, a fiber guided mode is established when its plane wave components totally reflect at the core–cladding boundary, and when ray trajectories are such that transverse resonance is established within the core. A ray propagates through a single round trip as it moves from a given starting point to a point at an increased distance in z at which it again achieves its starting orientation. Transverse resonance is established when the starting and finishing points of the ray path, having the same r and ϕ coordinates, have the same relative phase.

The condition for total reflection is established by considering the ray path as it reflects from a curved interface, as discussed in Section 2.7. A meridional ray will have no component in the direction of curvature (ϕ direction) and will thus experience total reflection when its incidence angle exceeds the critical angle. A skew ray, on the other hand, has a propagation component in ϕ, characterized by propagation constant $\beta_\phi = q/r$. At a critical radius within the cladding, the ϕ-directed phase velocity will thus exceed that of light in the cladding, and so a mode with only a ϕ component would radiate away. It is the z component of the ray that allows mode confinement. The net propagation constant for a confined wave will by definition have no radial component in the cladding. The wavevector magnitude in the cladding will therefore be $k_2 = (\beta_\phi^2 + \beta^2)^{1/2}$. At infinite distances, β_ϕ approaches zero, which means that β must be greater than n_2k_0 to prevent the mode from radiating. The restriction on β is thus identical to that of the slab guide; that is, $n_2k_0 < \beta < n_1k_0$.

In the case of meridional rays, it is possible to have TE and TM field configurations. Skew rays, on the other hand, cannot have pure TE or TM polarizations. This is because the slanted plane wave will change its orientation on each reflection, making it impossible to maintain purely transverse electric or magnetic fields over the entire ray path. The mode fields associated with a skew ray will thus have z components of both **E** and **H**. For this reason, these modes are known as "hybrid" modes and are given the designation EH or HE.

3.2.   FIELD ANALYSIS OF THE STEP INDEX FIBER

Field analysis begins by assuming field solutions of the form

$$\mathbf{E} = \mathbf{E}_0(r, \phi) \exp(-j\beta z) \qquad \mathbf{H} = \mathbf{H}_0(r, \phi) \exp(-j\beta z) \qquad (3.4)$$

As in the slab guide, the z component of **E** is again found using Eq. (2.25), rewritten in cylindrical coordinates:

$$\nabla_t^2 E_{z1} + (n_1^2 k_0^2 - \beta^2)E_{z1} = 0 \qquad r \le a \qquad (3.5)$$

$$\nabla_t^2 E_{z2} + (n_2^2 k_0^2 - \beta^2) E_{z2} = 0 \qquad r \geq a \tag{3.6}$$

where $(n_1^2 k_0^2 - \beta^2) = \beta_{t1}^2$ and $(n_2^2 k_0^2 - \beta^2) = \beta_{t2}^2$. Assuming transverse variation in both r and ϕ, the wave equation becomes, in general,

$$\frac{\partial^2 E_z}{\partial r^2} + \frac{1}{r} \frac{\partial E_z}{\partial r} + \frac{1}{r^2} \frac{\partial^2 E_z}{\partial \phi^2} + \beta_t^2 E_z = 0 \tag{3.7}$$

It is assumed that the solution for E_z will be a discrete series of modes, each of which has separated dependences on r, ϕ, and z in product form:

$$E_z = \sum_i R_i(r) \Phi_i(\phi) \exp(-j\beta_i z) \tag{3.8}$$

Each term (mode) in the expansion must itself be a solution of (3.7). A single mode, $E_z = R\Phi \exp(-j\beta z)$, can thus be substituted into (3.7) to obtain

$$\frac{r^2}{R} \frac{d^2 R}{dr^2} + \frac{r}{R} \frac{dR}{dr} + r^2 \beta_t^2 = -\frac{1}{\Phi} \frac{d^2 \Phi}{d\phi^2} \tag{3.9}$$

The left-hand side of (3.9) depends only on r, whereas the right-hand side depends only on ϕ. Since r and ϕ vary independently, it must follow that each side of (3.9) must be equal to a constant. Defining this constant as q^2, (3.9) separates into the following two equations:

$$\frac{d^2 \Phi}{d\phi^2} + q^2 \Phi = 0 \tag{3.10}$$

and

$$\frac{d^2 R}{dr^2} + \frac{1}{r} \frac{dR}{dr} + \left(\beta_t^2 - \frac{q^2}{r^2} \right) R = 0 \tag{3.11}$$

Identifying the term q/r as β_ϕ, it follows that $\beta_t^2 - q^2/r^2 = \beta_r^2$. Solving (3.10) results in

$$\Phi(\phi) = \begin{cases} \cos(q\phi + \alpha) \\ \sin(q\phi + \alpha) \end{cases} \tag{3.12}$$

where α is a constant phase shift. Again it will be noted that q must be an integer, since it is required that the field be self-consistent on each rotation of ϕ through 2π. q is known as the angular or azimuthal mode number.

Equation (3.11) is a form of Bessel's equation. Its solution will be in terms of Bessel functions of the form

$$R(r) = \begin{cases} AJ_q(\beta_t r) + A'N_q(\beta_t r) & \beta_t \quad \text{real} \\ CK_q(|\beta_t|r) + C'I_q(|\beta_t|r) & \beta_t \quad \text{imaginary} \end{cases} \tag{3.13}$$

where J_q and N_q are ordinary Bessel functions of the first and second kind, and of order q, which apply to cases of real β_t. If β_t is imaginary, then the solution will consist of modified Bessel functions, K_q and I_q. Figures 3.3 and 3.4 depict the behavior of these functions. More detailed discussions of Bessel function properties are found in Appendix A.

The basic properties of a guided mode in the fiber should be similar to those found in the slab guide. In particular, it is expected that (1) the solution in the core will be oscillatory, exhibiting no singularities, and (2) the solution in the cladding will monotonically decrease as radius increases. The first condition suggests that the ordinary Bessel functions should apply in the core region, and so $\beta_{t1} = (n_1^2 k_0^2 - \beta^2)^{1/2}$ is required to be real. Only the J_q functions can be used, since the N_q functions are infinite at $r = 0$. The second condition indicates that the K_q functions would exhibit the proper variation in the cladding, with I_q ruled out. Therefore $\beta_{t2} = (n_2^2 k_0^2 - \beta^2)^{1/2}$ is required to be imaginary.

Normalized transverse propagation and attenuation constants are defined:

$$u = \beta_{t1}a = a(n_1^2 k_0^2 - \beta^2)^{1/2} \tag{3.14}$$

$$w = |\beta_{t2}|a = a(\beta^2 - n_2^2 k_0^2)^{1/2} \tag{3.15}$$

Setting $A' = C' = 0$ in (3.13), and using the $\sin(q\phi)$ dependence in (3.12) (with $\alpha = 0$), we find the complete solution for E_z:

$$E_z = \begin{cases} AJ_q(ur/a)\sin(q\phi)\exp(-j\beta z) & r \leq a \\ CK_q(wr/a)\sin(q\phi)\exp(-j\beta z) & r \geq a \end{cases} \tag{3.16}$$

The integer requirement on q imposed by (3.12) carries over to the radial solution as well, thus limiting the Bessel functions to those of integer order. In similar fashion, the wave equation for $\mathbf{H}$ can be solved for the z component, leading to

$$H_z = \begin{cases} BJ_q(ur/a)\cos(q\phi)\exp(-j\beta z) & r \leq a \\ DK_q(wr/a)\cos(q\phi)\exp(-j\beta z) & r \geq a \end{cases} \tag{3.17}$$

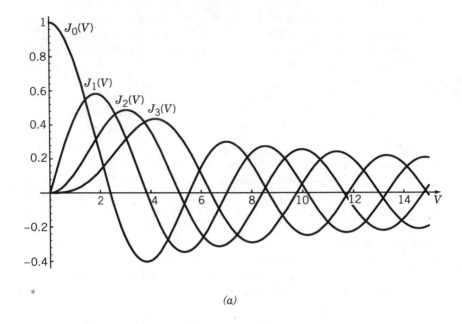

(a)

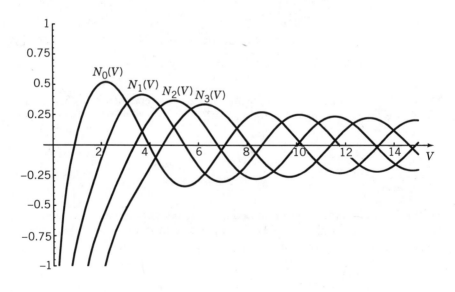

(b)

Figure 3.3. Ordinary Bessel functions.

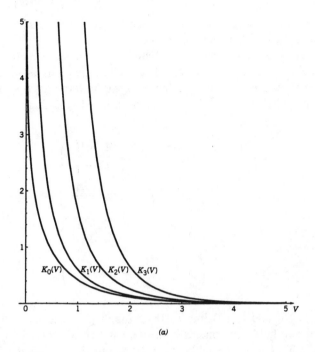

(a)

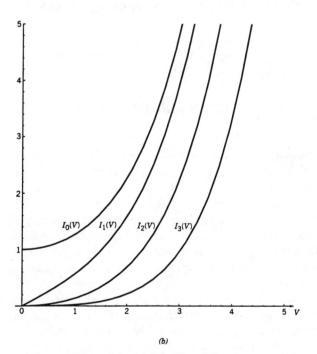

(b)

Figure 3.4. Modified Bessel functions.

It will be seen that choosing the cosine dependence for H_z enables continuity of all tangential field components at $r = a$.

The transverse field components are found using the same method that was used with the slab guide; specifically, Maxwell's equations are used to find general expressions for all transverse components in terms of first derivatives of the z components. In cylindrical coordinates, it is found that

$$E_r = \frac{-j}{\beta_t^2}\left(\beta\,\frac{\partial E_z}{\partial r} + \omega\mu\,\frac{1}{r}\,\frac{\partial H_z}{\partial \phi}\right) \tag{3.18}$$

$$E_\phi = \frac{-j}{\beta_t^2}\left(\beta\,\frac{1}{r}\,\frac{\partial E_z}{\partial \phi} - \omega\mu\,\frac{\partial H_z}{\partial r}\right) \tag{3.19}$$

$$H_r = \frac{-j}{\beta_t^2}\left(\beta\,\frac{\partial H_z}{\partial r} - \omega\epsilon\,\frac{1}{r}\,\frac{\partial E_z}{\partial \phi}\right) \tag{3.20}$$

$$H_\phi = \frac{-j}{\beta_t^2}\left(\beta\,\frac{1}{r}\,\frac{\partial H_z}{\partial \phi} + \omega\epsilon\,\frac{\partial E_z}{\partial r}\right) \tag{3.21}$$

Substituting (3.16) and (3.17) into (3.18) through (3.21) leads to the general expressions for all field components in the core and cladding regions. These are listed in Table 3.1 [1]. In the table, the primed functions, J_q' and K_q', indicate that their derivatives are taken with respect to the argument; for example,

$$J_q' = \frac{d}{d(ur/a)}\,J_q(ur/a).$$

3.3. MODE CLASSIFICATION AND THE EIGENVALUE EQUATIONS

The possible types of modes are understood by considering a few special cases. First, suppose that $q = 0$, and there is no z component of electric field ($A = C = 0$). In this case, Table 3.1 indicates that, in addition to H_z, the only nonzero field components will be E_ϕ and H_r. The *transverse electric* family of modes is thus obtained, designated TE_{0m}, where the subscript 0 indicates the value of the mode order, q; m is the mode rank, or radial mode number, which is greater than or equal to one. m has the same significance that it has in the slab guide, in which each value of m corresponds to one value each of k_{x1}, γ_2, and β. In the fiber a given m value is associated with a value each of u, w, and β. Transverse magnetic modes occur when $q = 0$ and when $H_z = 0$ ($B = D = 0$). Table 3.1 indicates that the only nonzero components in this case will be E_z, E_r, and H_ϕ. The subscripts in the designation TM_{0m} have the same meanings as in TE_{0m}. Modes of designation TE_{qm} and TM_{qm} where $q \neq 0$ are not possible. This is because too many field components would result, having too few constants

Table 3.1 General Step Index Fiber Fields in Phasor Form (exp $(-j\beta z)$ Dependence Understood)

$r < a$:

$$E_z = A J_q\left(\frac{ur}{a}\right) \sin(q\phi) \tag{3.22}$$

$$E_r = \left[-A \frac{j\beta}{(u/a)} J_q'\left(\frac{ur}{a}\right) + B \frac{j\omega\mu}{(u/a)^2} \frac{q}{r} J_q\left(\frac{ur}{a}\right)\right] \sin(q\phi) \tag{3.23}$$

$$E_\phi = \left[-A \frac{j\beta}{(u/a)^2} \frac{q}{r} J_q\left(\frac{ur}{a}\right) + B \frac{j\omega\mu}{(u/a)} J_q'\left(\frac{ur}{a}\right)\right] \cos(q\phi) \tag{3.24}$$

$$H_z = B J_q\left(\frac{ur}{a}\right) \cos(q\phi) \tag{3.25}$$

$$H_r = \left[A \frac{j\omega\epsilon_0 n_1^2}{(u/a)^2} \frac{q}{r} J_q\left(\frac{ur}{a}\right) - B \frac{j\beta}{(u/a)} J_q'\left(\frac{ur}{a}\right)\right] \cos(q\phi) \tag{3.26}$$

$$H_\phi = \left[-A \frac{j\omega\epsilon_0 n_1^2}{(u/a)} J_q'\left(\frac{ur}{a}\right) + B \frac{j\beta}{(u/a)^2} \frac{q}{r} J_q\left(\frac{ur}{a}\right)\right] \sin(q\phi) \tag{3.27}$$

$r > a$:

$$E_z = C K_q\left(\frac{wr}{a}\right) \sin(q\phi) \tag{3.28}$$

$$E_r = \left[C \frac{j\beta}{(w/a)} K_q'\left(\frac{wr}{a}\right) - D \frac{j\omega\mu}{(w/a)^2} \frac{q}{r} K_q\left(\frac{wr}{a}\right)\right] \sin(q\phi) \tag{3.29}$$

$$E_\phi = \left[C \frac{j\beta}{(w/a)^2} \frac{q}{r} K_q\left(\frac{wr}{a}\right) - D \frac{j\omega\mu}{(w/a)} K_q'\left(\frac{wr}{a}\right)\right] \cos(q\phi) \tag{3.30}$$

$$H_z = D K_q\left(\frac{wr}{a}\right) \cos(q\phi) \tag{3.31}$$

$$H_r = \left[-C \frac{j\omega\epsilon_0 n_2^2}{(w/a)^2} \frac{q}{r} K_q\left(\frac{wr}{a}\right) + D \frac{j\beta}{(w/a)} K_q'\left(\frac{wr}{a}\right)\right] \cos(q\phi) \tag{3.32}$$

$$H_\phi = \left[C \frac{j\omega\epsilon_0 n_2^2}{(w/a)} K_q'\left(\frac{wr}{a}\right) - D \frac{j\beta}{(w/a)^2} \frac{q}{r} K_q\left(\frac{wr}{a}\right)\right] \sin(q\phi) \tag{3.33}$$

to enable satisfaction of all boundary conditions. Physically, the $q \neq 0$ case corresponds to a skew ray, which cannot maintain purely transverse electric and magnetic field components, as described in Section 3.1.

For cases in which $q \neq 0$, hybrid modes result, having z components of both **E** and **H**. These are designated either HE_{qm} or EH_{qm}, depending on the characteristics of the eigenvalue equation. Modes with designations HE_{0m} and EH_{0m} are not possible, again because these fields would be too simplified to enable all boundary conditions to be satisfied.

With the transverse field components, eigenvalue equations can be derived that determine β, u, and w for the modes. Following the method of Okoshi [1], use is made of the requirement that all field components that are tangent to the core–cladding boundary at $r = a$ be continuous across it. All ϕ and z components apply, producing four equations in all. From Table 3.1, continuity of the z components of electric and magnetic fields results in

$$AJ_q(u) - CK_q(w) = 0 \qquad (3.34)$$

for $E_{z1}(a) = E_{z2}(a)$, and

$$BJ_q(u) - DK_q(w) = 0 \qquad (3.35)$$

for $H_{z1}(a) = H_{z2}(a)$. Continuity of the ϕ components at $r = a$ results in

$$A \frac{j\beta}{(u/a)^2} \frac{qJ_q(u)}{a} - B \frac{j\omega\mu}{(u/a)} J_q'(u) + C \frac{j\beta}{(w/a)^2} \frac{qK_q(w)}{a}$$
$$- D \frac{j\omega\mu}{(w/a)} K_q'(w) = 0 \qquad (3.36)$$

for E_ϕ, and

$$A \frac{j\omega\epsilon_0 n_1^2}{(u/a)} J_q'(u) - B \frac{j\beta}{(u/a)^2} \frac{qJ_q(u)}{a} + C \frac{j\omega\epsilon_0 n_2^2}{(w/a)} K_q'(w)$$
$$- D \frac{j\beta}{(w/a)^2} \frac{qK_q(w)}{a} = 0 \qquad (3.37)$$

for H_ϕ. Equations (3.34) through (3.37) can be written in matrix form as

$$[M] \cdot \begin{bmatrix} A \\ B \\ C \\ D \end{bmatrix} = 0 \qquad (3.38)$$

where $[M]$ is a matrix composed of the coefficients of A, B, C, and D in (3.34) through (3.37). For (3.38) to hold, the determinant of $[M]$ must vanish. Evaluating the determinant and requiring a zero result, we find [1]

$$\left(\frac{J_q'(u)}{uJ_q(u)} + \frac{K_q'(w)}{wK_q(w)} \right) \left(\frac{n_1^2}{n_2^2} \frac{J_q'(u)}{uJ_q(u)} + \frac{K_q'(w)}{wK_q(w)} \right)$$
$$= q^2 \left(\frac{1}{u^2} + \frac{1}{w^2} \right) \left(\frac{n_1^2}{n_2^2} \frac{1}{u^2} + \frac{1}{w^2} \right) \qquad (3.39)$$

Equation (3.39) is greatly simplified by the *weakly guiding approximation*, which states that $n_1 \approx n_2$. This condition is almost always satisfied in practical optical fibers. With the approximation, (3.39) becomes

$$\left(\frac{J'_q(u)}{uJ_q(u)} + \frac{K'_q(w)}{wK_q(w)} \right) = \pm q \left(\frac{1}{u^2} + \frac{1}{w^2} \right) \tag{3.40}$$

The case in which $q = 0$ applies to TE and TM modes. Using Bessel function identities, (3.40) becomes

$$u \frac{J_0(u)}{J_1(u)} = -w \frac{K_0(w)}{K_1(w)} \qquad TE_{0m}, TM_{0m} \tag{3.41}$$

For $q > 0$, the right-hand side of (3.40) will be either positive or negative. Equation (3.40) with positive right-hand side is the eigenvalue equation for EH_{qm} modes; with negative right-hand side, it is the equation for HE_{qm} modes.

The EH and HE designations originate from an early system in which comparisons are made of E_z and H_z to the transverse field components at some reference point [2]. This system is somewhat arbitrary, since results depend on the choice of the reference point. The EH or HE nomenclature is nevertheless convenient and descriptive in classifying modes according to the above cases that occur in (3.40).

For $q > 0$, (3.40) can be specialized for the two cases of positive and negative right-hand side by again using Bessel function identities. The results are

$$u \frac{J_q(u)}{J_{q+1}(u)} = -w \frac{K_q(w)}{K_{q+1}(w)} \qquad EH_{qm} \ (q \geq 1) \tag{3.42}$$

$$u \frac{J_q(u)}{J_{q-1}(u)} = w \frac{K_q(w)}{K_{q-1}(w)} \qquad HE_{qm} \ (q \geq 1) \tag{3.43}$$

As a further simplification, (3.41) to (3.43) can be written as a single equation by first rewriting (3.43) as

$$u \frac{J_{q-2}(u)}{J_{q-1}(u)} = -w \frac{K_{q-2}(w)}{K_{q-1}(w)} \qquad HE_{qm} \ (q \geq 1) \tag{3.44}$$

Defining a new integer variable, l, (3.41), (3.42), and (3.44) can be expressed as [1]

$$u \frac{J_{l-1}(u)}{J_l(u)} = -w \frac{K_{l-1}(w)}{K_l(w)} \qquad l = \begin{cases} 1 & TE_{0m}, TM_{0m} \\ q+1 & EH_{qm} \\ q-1 & HE_{qm} \end{cases} \tag{3.45}$$

The significance of (3.45) is that a set of degenerate EH, HE, TE, or TM modes will combine to form a new "composite" mode whose eigenvalue equation is (3.45). It will be shown in a later section that the electric field superposition of the combining modes is linearly polarized in the transverse plane; for this reason, the composite modes are referred to as the LP_{lm} modes. The LP system is a simplified way of describing modes in weakly guiding fibers. Furthermore, the LP modes are almost always observed in practice and are readily identified by their transverse intensity profiles.

3.4. SOLUTIONS OF THE EIGENVALUE EQUATIONS

A graphical solution method for the eigenvalue equations is demonstrated here for the cases of TE_{0m}, TM_{0m} (LP_{1m}), and HE_{1m} (LP_{0m}) modes [3]. To begin, the normalized frequency parameter, V, is defined as

$$V = (u^2 + w^2)^{1/2} = ak_0(n_1^2 - n_2^2)^{1/2} = n_1 a k_0 \sqrt{2\Delta} \qquad (3.46)$$

This important parameter, often called the V number, embodies the fiber structural parameters and frequency. It is used with (3.45) to determine cutoff conditions for the modes, propagation constants, and information on power confinement. It is seen that V is directly analogous to the radius function R, defined in (2.46), that is used in the graphical solution for the slab waveguide.

Consider first (3.41), which can be rewritten to read

$$\frac{J_1(u)}{J_0(u)} = -\frac{u}{w}\frac{K_1(w)}{K_0(w)} \qquad (3.47)$$

Noting the behavior of J_0 and J_1 as shown in Fig. 3.3, it can be seen that the left-hand side of (3.47), when plotted as a function of u, will produce curves that resemble those of a tangent function. These are shown in the two graphs that comprise Fig. 3.5. In the graphs, the asymptotes and zeros correspond to appropriate zeros in Fig. 3.3. The right-hand side of (3.47) is plotted as the curve shown in the lower half-plane of each graph. Its intersections with the J_1/J_0 curves indicate solutions to (3.47). The values of u thus obtained can be used to find w and β, knowing V. The first plot was made using $V = 2$. From (3.46), the maximum value of u for a given V occurs when $w = 0$, at which $u = V$. A vertical line is drawn on the first graph at $u = 2$ that marks this position; it thus indicates the upper bound on the solutions for u that will appear on the graph. The second graph was plotted for $V = 8$, showing the vertical boundary at $u = 8$.

HE_{1m} (LP_{0m}) modes can be exhibited on the same graph by first rewriting

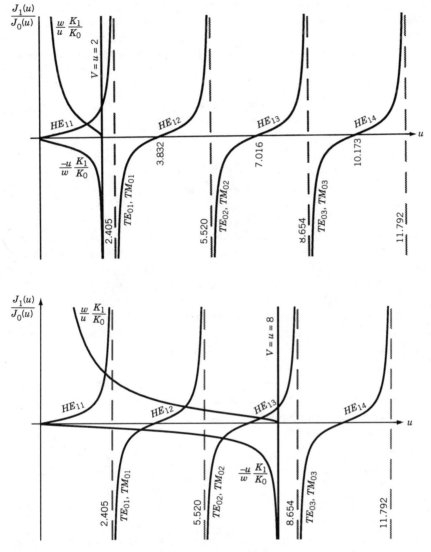

Figure 3.5. Graphical solutions of Eqs. (3.47) and (3.48) for $V = 2$ (upper) and $V = 8$ (lower).

(3.43) to read

$$\frac{J_1(u)}{J_0(u)} = \frac{w}{u}\frac{K_1(w)}{K_0(w)} \tag{3.48}$$

The left-hand side of (3.48) produces the same curves as the left-hand side of (3.47). The right-hand side of (3.48) produces the curve having negative slope shown in the upper half-plane of each graph in Fig. 3.5. The fact that this curve approaches infinity as u becomes small is evident from the equation and by inspection of Fig. 3.4. The fact that the function reaches zero at $V = u$ is not obvious but can be understood by noting the behavior of the modified Bessel functions for small w, where

$$\frac{w}{u} \frac{K_1(w)}{K_0(w)} \approx \frac{1}{u \ln (2/1.8w)} \tag{3.49}$$

The above approaches zero as w approaches zero.

The branches of $J_1(u)/J_0(u)$ in Fig. 3.5 are labeled by their mode designations, where the values of m on a given branch will differ for the two mode types. A change in frequency or numerical aperture will change V and will thus change the position of the vertical "V line" on the graph accordingly. The curves that correspond to the right-hand sides of (3.47) and (3.48) either stretch or contract with the vertical line position, thus producing fewer or more solutions, depending on whether V decreases or increases.

A few important points should be noted.

1. If the vertical line is positioned at a zero of a $J_1(u)/J_0(u)$ curve, then the cutoff point is reached for the HE_{1m} (LP_{0m}) mode associated with that branch.

2. If the vertical line is positioned at a lower asymptote of a $J_1(u)/J_0(u)$ curve, then the cutoff point is reached for the TE_{0m} or TM_{0m} (LP_{1m}) mode corresponding to that branch number.

3. As V increases, all modes having $u < V$ will continue to have solutions and will thus continue to propagate.

4. There is no cutoff for the HE_{11} (LP_{01}) mode. The fiber thus propagates this mode only for values of V in the range $0 < V < 2.405$. The upper graph of Fig. 3.5 shows a case of single-mode operation.

5. Values of w for a given mode can achieve any value between zero and infinity.

6. The values of u are bounded for a given mode. This fact can be deduced from the graph or from (3.45) by noting that, at cutoff ($w = 0$), u will be the mth zero of J_{l-1} (this does not include the zero at $u = 0$, except for the case when $l = 0$). Far above cutoff, as $w \to \infty$, u will approach the mth zero of J_l. For example, u values for the TE_{01} mode occur within the range $2.405 \leq u \leq 3.832$.

7. Values of V for a given mode will occur over the range $V_c \leq V \leq \infty$, where V_c is the value of V at cutoff for that mode.

Similar plots can be made for HE_{qm} modes for $q \geq 2$, and for EH_{qm} modes. These will produce curves of J_l/J_{l-1} that will fall between some of the curves shown in the graphs, making the progression of mode appearance with increasing V somewhat more complicated than is depicted there. Nevertheless, single-mode operation will still occur over the range $0 < V < 2.405$.

3.5. CUTOFF CONDITIONS

Cutoff for a given mode can be determined directly from (3.45) by setting $w = 0$, which means that $u = V$. Under this condition, the cutoff condition is obtained from (3.45):

$$\frac{V J_{l-1}(V)}{J_l(V)} = 0 \tag{3.50}$$

The above will be satisfied for all values of V that correspond to zeros of J_{l-1}. The special case of $V = 0$ is handled by the small argument approximation of (3.50):

$$\frac{V J_{l-1}(V)}{J_l(V)} \approx \frac{l V(V/2)^{l-1}}{(V/2)^l} = 2l \tag{3.51}$$

The above will be zero only when $l = 0$. Hence, as V approaches zero, (3.50) is satisfied only by the special case

$$\frac{V J_{-1}(V)}{J_0(V)} = 0 \tag{3.52}$$

where $J_{-1}(V) = -J_1(V)$. Equation (3.52) thus applies to the HE_{11} (LP_{01}) mode, having cutoff at $V = 0$.

For the remaining zeros, a simplified form of (3.50) is used:

$$J_{l-1}(V) = 0 \tag{3.53}$$

It will be recalled that $l = 1$ for TE and TM modes, $q + 1$ for EH modes, and $q - 1$ for HE modes. Using (3.53), each Bessel function zero is considered, and the modes with cutoff occurring at that value of V are established. For example, 2.405 is the first zero of J_0. Equation (3.53) becomes $J_0(2.405) = 0$, which indicates that $l = 1$. Considering the above relations between l and q, the modes TE_{01}, TM_{01}, and HE_{21} will all achieve cutoff at $V = 2.405$. The value of m is one for these modes because $V = 2.405$ is the first instance in which any of these modes (with the given value of q) occur as V increases. From (3.45), it is seen that these modes will all have the same eigenvalue equation, which means that they will be *degenerate*; that is, they will all have the same value of β. This is true in the weakly guiding approximation, despite the fact that the field configurations for these modes differ significantly from one another. In general, any modes that have the same l and same m will be degenerate.

3.6. SPECIFIC FIELD FUNCTIONS

Specific formulas for the fields of the four types of modes can be derived using Table 3.1, along with the results of Section 3.3. To begin, (3.34) and (3.35) are used to enable (3.36) to be expressed as

$$
j \, \frac{\beta q A}{a} \left[\frac{1}{(u/a)^2} + \frac{1}{(w/a)^2} \right] J_q(u)
$$
$$
- j\omega\mu B \left[\frac{1}{(u/a)^2} J_q'(u) + \frac{1}{(w/a)^2} \frac{J_q(u)}{K_q(w)} K_q'(w) \right] = 0 \quad (3.54)
$$

The Bessel function derivatives can be expressed in nondifferentiated form using identities (A.20) and (A.31) in Appendix A. Equation (3.54) then becomes

$$
j\beta q A \left[\frac{1}{(u/a)^2} + \frac{1}{(w/a)^2} \right]
$$
$$
+ j\omega\mu B \left\{ a^2 \left[\frac{1}{u} \frac{J_{q+1}(u)}{J_q(u)} + \frac{1}{w} \frac{K_{q+1}(w)}{K_q(w)} \right] \right.
$$
$$
\left. - q \left[\frac{1}{(u/a)^2} + \frac{1}{(w/a)^2} \right] \right\} = 0 \quad (3.55)
$$

Invoking (3.42) simplifies the above to yield the relation between coefficients A and B for EH modes. Beginning again from (3.54), identities (A.21) and (A.32) could instead have been used, along with (3.43), to yield the relation between A and B for HE modes. These results are summarized as

$$
B = \pm \frac{\beta A}{\omega\mu} \quad (3.56)
$$

where the positive and negative signs apply to EH and HE modes, respectively. Equations (3.34) and (3.35) are then used, along with (3.56), to yield expressions for C and D in terms of A:

$$
D = \pm \frac{\beta A}{\omega\mu} \frac{J_q(u)}{K_q(w)} = \pm \frac{\beta C}{\omega\mu} \quad (3.57)
$$

Use of (3.56) and (3.57) in the field expressions of Table 3.1, along with Bessel function identities (A.20), (A.21), and (A.29) to (A.32), leads to the following electric field functions for the EH and HE modes, under the weakly guiding approximation, with $E_0 \equiv j\beta Aa/u$ and with $\exp(-j\beta z)$ understood:

$$EH_{qm}: \quad E_r(r \leq a) = E_0 J_{q+1}\left(\frac{ur}{a}\right) \sin(q\phi) \tag{3.58}$$

$$E_r(r \geq a) = -E_0 \frac{u}{w} \frac{J_q(u)}{K_q(w)} K_{q+1}\left(\frac{wr}{a}\right) \sin(q\phi) \tag{3.59}$$

$$E_\phi(r \leq a) = -E_0 J_{q+1}\left(\frac{ur}{a}\right) \cos(q\phi) \tag{3.60}$$

$$E_\phi(r \geq a) = E_0 \frac{u}{w} \frac{J_q(u)}{K_q(w)} K_{q+1}\left(\frac{wr}{a}\right) \cos(q\phi) \tag{3.61}$$

$$HE_{qm}: \quad E_r(r \leq a) = -E_0 J_{q-1}\left(\frac{ur}{a}\right) \sin(q\phi) \tag{3.62}$$

$$E_r(r \geq a) = -E_0 \frac{u}{w} \frac{J_q(u)}{K_q(w)} K_{q-1}\left(\frac{wr}{a}\right) \sin(q\phi) \tag{3.63}$$

$$E_\phi(r \leq a) = -E_0 J_{q-1}\left(\frac{ur}{a}\right) \cos(q\phi) \tag{3.64}$$

$$E_\phi(r \leq a) = -E_0 \frac{u}{w} \frac{J_q(u)}{K_q(w)} K_{q-1}\left(\frac{wr}{a}\right) \cos(q\phi) \tag{3.65}$$

The TE and TM electric fields in the transverse plane are found by setting $q = 0$ in the equations of Table 3.1, and setting all sinusoidal functions to unity. The results are expressed as follows:

$$TE_{0m}, TM_{0m}: \quad E_t(r \leq a) = E_0 J_1\left(\frac{ur}{a}\right) \tag{3.66}$$

$$E_t(r \geq a) = -E_0 \frac{u}{w} \frac{J_0(u)}{K_0(w)} K_1\left(\frac{wr}{a}\right) \tag{3.67}$$

where $E_t = E_\phi$ for TE modes and $E_t = E_r$ for TM. The magnetic fields for all the modes can be found using a similar procedure. As an example, Fig. 3.6 shows the appearance of the electric and magnetic fields of the TE_{02} mode within the core. It is evident that the electric field appears as a series of concentric circles in the transverse plane and thus exhibits no variation with ϕ; this would be determined by the fact that $q = 0$. It is also evident that the fields go through two half-cycles in the radial direction; the number of half-cycles is determined by the value of m, which is two in this case. The mode as pictured is at a frequency much higher than the cutoff frequency for TE_{02}. Consequently, the

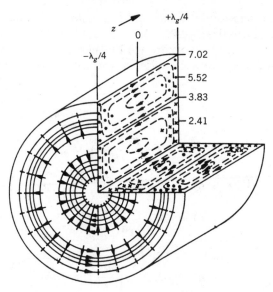

Figure 3.6. Field plots of **E** (solid lines) and **H** (dashed lines) for the TE_{02} mode in the core of a step index fiber. The length shown in one-half of a guide wavelength, where $\lambda_g = 2\pi/\beta$. The core radius is indicated in units of ur/a, and the mode is shown for $V \gg V_c$ [4].

value of u for this mode approaches 7.016. This is seen from the graphical solution of Fig. 3.5, as $V \to \infty$.

3.7. LP MODES

An alternate designation system is used for degenerate sets of modes in the weakly guiding approximation. This is the LP_{lm} notation, where LP means "linearly polarized." The name originates from the fact that by superimposing transverse fields of modes in a degenerate set, a resultant field can be obtained that is polarized in only one cartesian direction. Another way of motivating this concept is to consider the effect of coupling linearly polarized light into a weakly guiding fiber. In an ideal fiber, the light will maintain its linear polarization, and so degenerate modes must superimpose in such a way to accomplish this.

The LP modes can be identified by using the cutoff condition (3.53). This function is satisfied for $V = 2.405$ for $l = 1$ and $m = 1$. The corresponding degenerate modes, as described in Section 3.5, are TE_{01}, TM_{01}, and HE_{21}. These together form the LP_{11} set. The next Bessel function zero (enabling satisfaction of (3.53)) will be 3.832. This applies to the J_1 function, which, from (3.53), means that $l = 2$ or $l = 0$ (the latter arises from the possibility of J_{-1}, which is equal to $-J_1$). The case of $l = 0$ applies to an HE mode for which $q = 1$. The HE_{12} mode is thus obtained. The value of m is 2, because this is the second occurrence of the HE_{1m} mode type. In the new notation, this mode is called LP_{02}. For $l = 2$, the HE_{31} and EH_{11} modes result. Together, these modes form the LP_{21} set. Considering the second zero of J_0 (5.520), it is found that $l = 1$, leading to the modes TM_{02}, TE_{02}, and HE_{22}, which together form the LP_{12} set. By continuing in this fashion (and including the higher order Bessel functions), the progression of modes can be established as far as desired. The first 12 LP modes, along with relevant parameters, are listed in Table 3.2.

The LP modes can in fact be obtained directly from the wave equation by postulating solutions having the linearly polarized form, as described later in this section, and as performed in [5]. They can also be obtained by simple additions of the mode functions of Section 3.6. Beginning with LP_{1m} case, it is evident from Table 3.2 that these modes will always be associated with HE_{2m}, TE_{0m}, and TM_{0m} component modes. The various ways in which the latter modes can be added will yield the characteristic LP_{11} field pattern having four possible orientations and polarizations as illustrated in Fig. 3.7. For example, consider subtracting the TE_{0m} mode field from that of the HE_{2m} mode. In the core, this is done by subtracting (3.66) (E_ϕ) from (3.64) (for $q = 2$) and then using that result, along with (3.62), in the coordinate transformation, $E_x = E_r \cos \phi - E_\phi \sin \phi$, to yield the linearly polarized field:

$$E_x = 2E_0 J_1 \left(\frac{ur}{a} \right) \sin \phi \qquad r \le a \qquad (3.68)$$

By using the transformation $E_y = E_r \sin \phi + E_\phi \cos \phi$, E_y is found to be zero. In the cladding, a similar procedure is used by subtracting (3.67) from (3.65),

Table 3.2 Cutoff Conditions and Designations of the First 12 LP Modes in a Step Index Fiber

V_c	Bessel Function	l	Degenerate Modes	LP Designation
0	—	0	HE_{11}	LP_{01}
2.405	J_0	1	$TE_{01}, TM_{01}, HE_{21}$	LP_{11}
3.832	J_1	2	EH_{11}, HE_{31}	LP_{21}
3.832	J_{-1}	0	HE_{12}	LP_{02}
5.136	J_2	3	EH_{21}, HE_{41}	LP_{31}
5.520	J_0	1	$TE_{02}, TM_{02}, HE_{22}$	LP_{12}
6.380	J_3	4	EH_{31}, HE_{51}	LP_{41}
7.016	J_1	2	EH_{12}, HE_{32}	LP_{22}
7.016	J_{-1}	0	HE_{13}	LP_{03}
7.588	J_4	5	EH_{41}, HE_{61}	LP_{51}
8.417	J_2	3	EH_{22}, HE_{42}	LP_{32}
8.654	J_0	1	$TE_{03}, TM_{03}, HE_{23}$	LP_{13}

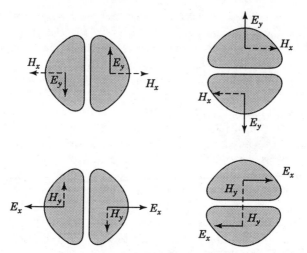

Figure 3.7. LP_{11} field patterns, showing the four possible configurations. (Adapted from ref. 5.)

and then using (3.41) and (3.43) to obtain

$$E_x = 2E_0 \frac{J_1(u)}{K_1(w)} K_1\left(\frac{wr}{a}\right) \sin\phi \qquad r \geq a \qquad (3.69)$$

LP modes having l values of 2 and higher will be formed by combinations of $HE_{l+1,m}$ and $EH_{l-1,m}$ modes. The resultant fields in the core are determined by first setting $q = l + 1$ in (3.62) and in (3.64), and $q = l - 1$ in (3.58) and in (3.60). Then, for example, (3.62) can be subtracted from (3.58), and (3.64) from (3.60). Using the above coordinate transformations, the resulting cartesian fields will be $E_y = 0$ and

$$E_x = 2E_0 J_l\left(\frac{ur}{a}\right) \sin(l\phi) \qquad r \leq a \qquad (3.70)$$

In the cladding, a similar procedure yields

$$E_x = 2E_0 \frac{J_l(u)}{K_l(w)} K_l\left(\frac{wr}{a}\right) \sin(l\phi) \qquad r \geq a \qquad (3.71)$$

Finally, for the LP_{0m} case, only the HE_{1m} mode is present. It is left as an exercise (Problem 3.5) to show that the resulting cartesian fields (either E_x or E_y) will be

$$E_{LP_{0m}} = E_0 J_0\left(\frac{ur}{a}\right) \qquad r \leq a \tag{3.72}$$

$$E_{LP_{0m}} = E_0 \frac{J_0(u)}{K_0(w)} K_0\left(\frac{wr}{a}\right) \qquad r \geq a \tag{3.73}$$

Figure 3.8 depicts the field configuration for the LP_{01} HE_{11}) mode, showing the linearly polarized appearance of the fields in the transverse plane.

Experimentally, the LP modes are observed as intensity patterns (proportional to EE^*) in the transverse plane. By using (3.68) to (3.73), the intensity functions in the core and cladding for any LP mode can be expressed as

$$I_{lm} = I_0 J_l^2\left(\frac{ur}{a}\right) \sin^2(l\phi) \qquad r \leq a \tag{3.74}$$

$$I_{lm} = I_0 \left(\frac{J_l(u)}{K_l(w)}\right)^2 K_l^2\left(\frac{wr}{a}\right) \sin^2(l\phi) \qquad r \geq a \tag{3.75}$$

where I_0 is the peak intensity. Note that setting $l = 0$ for the LP_{0m} case has the effect of eliminating the sine function (not reducing it to zero). This is because the angular orientation of the cartesian axes is purely arbitrary in a round fiber, thus enabling an arbitrary phase to be added to the argument of the sine without changing the basic result. The effect of increasing the radial mode order, m, is to increase the value of u for a given V, as was demonstrated in the graphical solution of the eigenvalue equations. The result is that for higher m, more radial oscillations will be observed in the mode field and intensity patterns.

The intensity patterns for the first six modes, calculated from (3.74) and (3.75), are shown in Fig. 3.9. As observed in the figure, the forms of the patterns

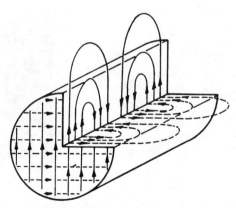

Figure 3.8. Plots of **E** (solid lines) and **H** (dashed lines) for the LP_{01} mode in the core of a step index fiber, showing the linearly polarized nature of the transverse fields [6].

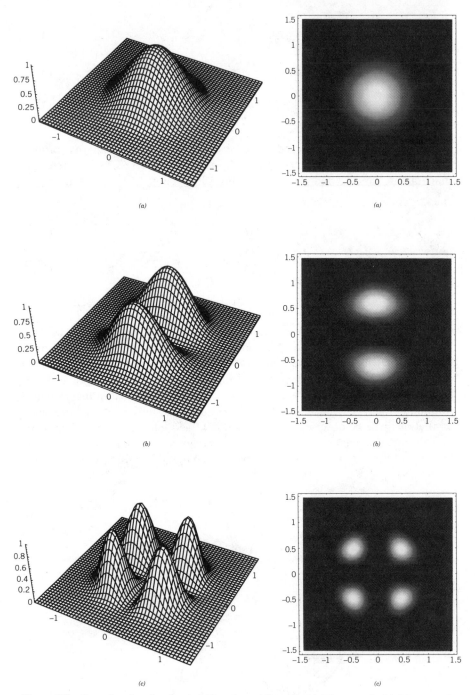

Figure 3.9. Intensity plots for the six LP modes, with $a = 1$. (a) LP_{01}: $u = 2$. (b) LP_{11}: $u = 3$. (c) LP_{21}: $u = 4.5$. (d) LP_{02}: $u = 4.5$. (e) LP_{31}: $u = 5.6$. (f) LP_{12}: $u = 6.3$.

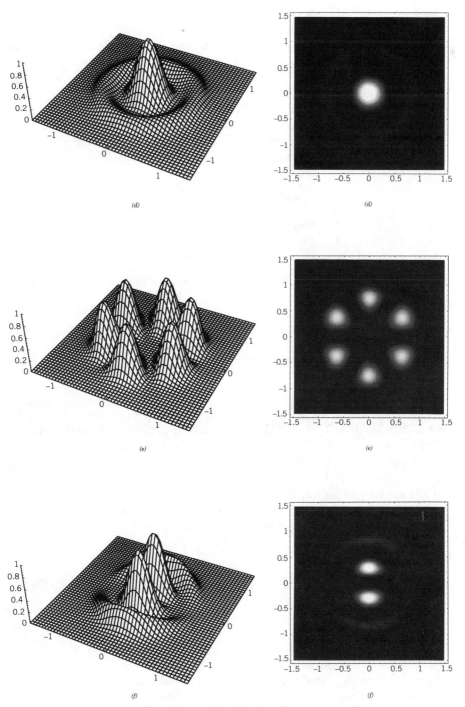

(d)   *(d)*

(e)   *(e)*

(f)   *(f)*

Figure 3.9. (*Continued*)

can be deduced from the subscripts l and m. The value of l is one-half the number of minima (or maxima) that occur in the intensity pattern as ϕ varies through 2π radians; m is the number of maxima in the intensity pattern that occur along a radial line between zero and infinity.

The propagation constant for each LP mode can be found by solving (3.45) either numerically or with the aid of the approximation $w \approx (V^2 - u_c^2)^{1/2}$, where u_c is the value of u at cutoff for the mode in question. For mode LP_{lm}, the latter is equal to the mth zero of J_{l-1} (excluding the zero root, except for the case of LP_{0m}). The analytic procedure, outlined in ref. 5, yields the normalized transverse propagation constant as a function of V:

$$u(V) \approx u_c \exp \left(\frac{\sin^{-1}(s/u_c) - \sin^{-1}(s/V)}{s} \right) \tag{3.76}$$

where $s \equiv (u_c^2 - l^2 - 1)^{1/2}$. For the LP_{01} mode, a more careful approximation yields [5]:

$$u_{01}(V) \approx \frac{(1+\sqrt{2})V}{1 + (4 + V^4)^{1/4}} \tag{3.77}$$

A test of the above formulas can be made by noting the value of u obtained for a given mode as $V \to \infty$. These values can be obtained directly from (3.45) as the mth zero of J_l. The values obtained from (3.76) and (3.77) are within 2% of the correct values, thus providing an indication of their general accuracy over the full range of V [5].

With $u(V)$ determined, the propagation constant for a given LP mode, β_{lm}, can be determined using (3.14). A useful parameter for this evaluation is the normalized propagation constant, defined as

$$b = 1 - \frac{u^2}{V^2} = \frac{\beta^2 - n_2^2 k_0^2}{n_1^2 k_0^2 - n_2^2 k_0^2} \approx \frac{n_{\text{eff}} - n_2}{n_1 - n_2} \tag{3.78}$$

where the latter approximation is valid in the case where $n_1 \approx n_2$. The *effective guide index* is defined as $n_{\text{eff}} \equiv \beta/k_0$. It is seen that b varies within the range $0 < b < 1$, achieving a value of zero at cutoff ($w = 0$) and unity far above cutoff (V and w both approach infinity while u is bounded). The mode propagation constant is thus

$$\beta_{lm} = n_2 k_0 (1 + 2b\Delta)^{1/2} \approx n_2 k_0 (1 + b\Delta) = n_2 k_0 [1 + \Delta(1 - u_{lm}^2/V^2)]$$

$$(3.79)$$

Figure 3.10 shows plots of b, calculated using (3.76) and (3.77) in (3.78), for several LP modes as functions of V. As plotted on this scale, these results are essentially indistinguishable from those given by the exact numerical solution [5].

When considering single-mode fibers, the accuracy of the LP_{01} curve in Fig. 3.10 is of increased importance. It was in fact found to be accurate to within 5% over the range of V between 2.0 and 3.0, with the error increasing to around 10% as V decreases to 1.5 [7]. Numerous other approximate formulas exist as alternatives to (3.77) for determining b. The best of these was found by Rudolf and Neumann [8], who recognized that w is a nearly linear function of V over the range $1.3 < V < 3.5$. They were thus able to approximate w over this range by the simple function

$$w \approx 1.1428V - 0.9960 \qquad (3.80)$$

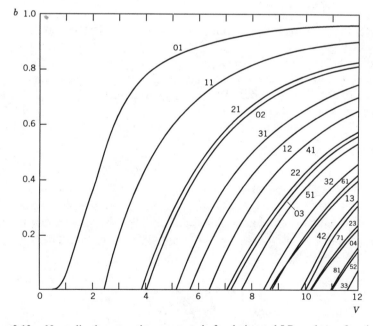

Figure 3.10. Normalized propagation constant, b, for designated LP modes as functions of V. (Adapted from ref. 5.)

from which

$$b \approx \frac{(1.1428V - 0.9960)^2}{V^2} \tag{3.81}$$

The error in the approximation in (3.81) is less than 0.1% over the important range of $1.5 < V < 2.4$ [8]. Precise knowledge of b and its variation with V is important in characterizing fiber dispersion, as will be considered in Chapter 5.

It is now possible to examine some of the implications of the LP modes more closely. To begin, note that under the weakly guiding approximation, the ray trajectories that are associated with the various modes will deviate very slightly from the z direction. As a consequence, the mode fields will have z components that are essentially negligible, thus enabling the treatment of the LP modes approximately as nonuniform plane waves. These propagate through the fiber medium, which is in turn characterized by the mode-dependent effective index, n_{eff}. Since the transverse fields for a given mode will assume only one cartesian direction, the wave equation, (2.25), is fully separable, thus enabling (3.5) and (3.6) to be written in terms of E_x, for example (although E_x is still expressed as a function of cylindrical coordinate variables). The steps in solving (3.5) and (3.6), thus modified, are as described previously and lead to the form of E_x given by (3.16), but with one change: the integer q in (3.16) (for E_x) must be replaced with the LP mode number, l, such that the solution is consistent with (3.70) through (3.73). This means that β_ϕ for the LP modes will be given by l/r; modes of designation LP_{0m} are thus identified as meridional. In other words, HE_{1m} modes, associated with skew rays in general, become meridional in the weakly guiding case.

A further consequence of the vanishing z components is that the electric and magnetic fields for the LP modes are simply related through the intrinsic impedance of the fiber material. That is, $E_x \approx \eta_2 H_y$ and $E_y \approx -\eta_2 H_x$, where $\eta_2 = \eta_0/n_2$ (assuming $n_1 \approx n_2$). Thus the magnetic fields associated with (3.70) through (3.73) are obtained by replacing E_0 with E_0/η_2 (with the appropriate sign).

3.8. POWER CONFINEMENT

The extent to which a propagating mode is confined to the fiber can be measured by the ratio of the power carried in the cladding to the total power that propagates in the mode:

$$\nu = \frac{P_{\text{clad}}}{P_{\text{core}} + P_{\text{clad}}} \tag{3.82}$$

The two power quantities involved are found by integrating the z component of the Poynting vector over the cross section of each region, as was done for the slab waveguide in Section 2.4. For the LP modes, the core and cladding powers are found through

$$P_{core} = \frac{1}{2} \int_0^{2\pi} \int_0^a \text{Re}\,\{E_{x1}H_{y1}^* - E_{y1}H_{x1}^*\}r\,dr\,d\phi \tag{3.83}$$

$$P_{clad} = \frac{1}{2} \int_0^{2\pi} \int_a^\infty \text{Re}\,\{E_{x2}H_{y2}^* - E_{y2}H_{x2}^*\}r\,dr\,d\phi \tag{3.84}$$

where E_x, E_y, H_x, and H_y are expressed as functions of r and ϕ. Using (3.83), (3.84), and the mode field expressions, we find that, in general [5],

$$\nu = \left(\frac{u^2}{V^2}\right)\left(1 - \frac{K_l^2(w)}{K_{l+1}(w)K_{l-1}(w)}\right) \tag{3.85}$$

Figure 3.11 shows calculated values of ν and $1 - \nu$ (the fraction of the total mode power that resides in the core) for various LP modes. Interestingly, modes

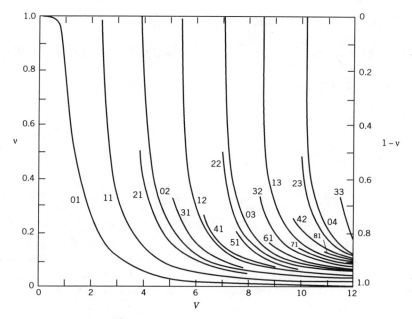

Figure 3.11. Power confinement of indicated LP modes as functions of V. (Adapted from ref. 5.)

for which $l = 0$ and $l = 1$ have essentially zero power in the core at cutoff. Modes having $l \geq 2$, however, exhibit appreciable power in the core at cutoff according to the ratio $P_{\text{core}}/P_{\text{clad}} = l - 1$. It is evident that the power in all modes becomes more concentrated in the core as V increases. This is a desirable feature, since with tighter power confinement, the fiber is less susceptible to losses associated with bending, as will be shown in Chapter 4. On the other hand, tighter confinement in the single-mode case increases the difficulty of coupling light into the fiber, since more precise alignment is needed between the input beam and the fiber core; this could lead to increased coupling losses. Consequently, some compromise is usually adopted in practice.

An additional fact concerning power flow in the fiber waveguide is that the modes are orthogonal, as is true in any waveguide. This property is embodied in the following relation, analogous to Eq. (2.54):

$$\int_0^{2\pi} \int_0^{\infty} \text{Re}\{\hat{\mathbf{E}}_{\alpha\beta} \times \hat{\mathbf{H}}_{\gamma\delta}\} \cdot \hat{\mathbf{a}}_z r \, dr \, d\phi = \delta_{\alpha\gamma} \delta_{\beta\delta} \qquad (3.86)$$

where $\hat{\mathbf{E}}_{\alpha\beta}$ and $\hat{\mathbf{H}}_{\gamma\delta}$ are mode fields having appropriately normalized amplitudes. α and γ specify values of l, and β and δ are values of m. Guided power flow is thus restricted to occur within the individual modes; no other transport mechanism exists.

3.9. CUTOFF WAVELENGTH

The cutoff wavelength for any mode is defined as the maximum wavelength at which that mode will propagate. It is the value of λ that corresponds to V_c for the mode in question. For each LP mode, the two parameters are related through

$$\lambda_c(lm) = \frac{2\pi a}{V_c(lm)} \sqrt{n_1^2 - n_2^2} = \frac{\lambda V}{V_c(lm)} \qquad (3.87)$$

The range of wavelengths over which mode lm will propagate is thus $0 < \lambda < \lambda_c(lm)$.

For a fiber to operate single mode, the operating wavelength must be greater than the cutoff wavelength for the LP_{11} mode. This latter quantity is in fact an important specification for a single-mode fiber and is usually given the designation λ_c (rather than $\lambda_c(11)$). λ_c is found from (3.87) by setting $V_c = 2.405$. The range of wavelengths for single-mode operation is thus $\lambda > \lambda_c$.

In situations involving single-mode fiber where the operating wavelength is fixed, it is often more convenient to specify λ_c, rather than V, since it is easier to visualize and use comparisons between wavelengths, and since wavelength is a directly measurable quantity. The importance of measurement becomes clear when realizing that (3.87) is valid for a perfectly straight fiber, in which the refractive indices and core radius are precisely known. The fact that this is not the case in practice means that measurement methods and the associated standards are needed to adequately characterize single-mode fiber.

The single-bend attenuation method has in recent years become a standard technique for measuring λ_c. The apparatus is shown in Fig. 3.12a. The light that is coupled into the fiber is tunable over a wavelength range that allows single-mode or bimode operation. In the latter case, the optical power will be distributed between the LP_{01} and LP_{11} modes. The experiment is designed to measure the wavelength at which an acceptably low fraction of the total power is carried in the LP_{11} mode. In the vicinity of λ_c, LP_{11}, being very weakly confined, is susceptible to loss arising from bends in the fiber, in addition to fiber length. It is thus necessary to establish a standard length to be used, as well as a standard radius at which the fiber is wound. The accepted standard is a fiber length of 2 m, loosely wound in a loop of 14 cm radius. The first stage of the measurement is as shown in step 1 in Fig. 3.12a. The large spot launch assures the excitation of both LP_{01} and LP_{11} modes. As the light frequency is varied, the detected power, $P_1(\lambda)$, will be a constant. In step 2, a relatively tight bend is placed in the fiber. The bend acts to increase the loss of LP_{11} (if it is propagating), such that the overall detected power, $P_2(\lambda)$, will be decreased under bimode operation. The loss arising from the bend is calculated by comparing the detected powers from steps 1 and 2 as

$$\text{Loss} = 10 \log_{10} \frac{P_1(\lambda)}{P_2(\lambda)} \tag{3.88}$$

The step 2 experiment is performed, and the loss is calculated as the wavelength is tuned. A typical result is shown in Fig. 3.12b. At wavelengths greater than the cutoff wavelength, the loss is essentially zero, as would be expected since only the LP_{01} mode propagates. A sharp increase in loss is seen as the wavelength is decreased, arising from the onset (and loss) of power in LP_{11}. The cutoff wavelength is defined as the value of λ at which the loss has increased to 0.1 dB above the zero level. This corresponds to about 2% of the total power propagating in the LP_{11} mode and implies an LP_{11} loss of about 10 dB/m.

3.10. GAUSSIAN APPROXIMATION FOR THE LP_{01} MODE FIELD

Previously, it has been found that a number of simplifications in the mode descriptions have resulted from the weakly guiding approximation. As an added example of this, consider the situation in which incident light in the form of a gaussian beam (from a well-collimated laser source) couples into a single-mode fiber through normal incidence at one end. Since the fiber is weakly guiding

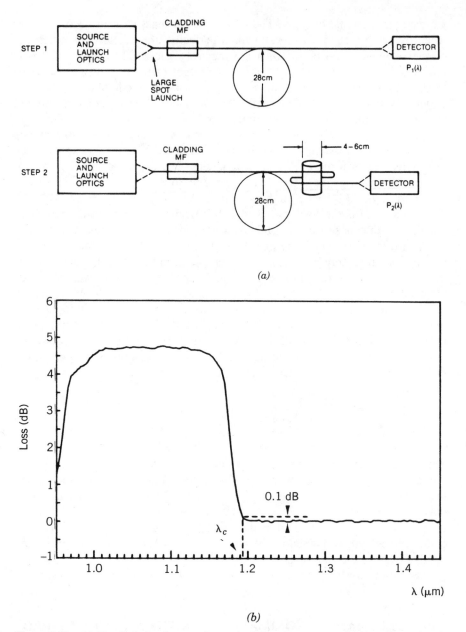

(a)

(b)

Figure 3.12. (a) The single-bend attenuation method, for measurement of the cutoff wavelength. (b) Typical power loss plot, obtained from the measurement apparatus of Fig. 3.12a [9].

$(n_1 \approx n_2)$, one would suspect intuitively that the fiber structure will do little to change the transverse profile of the beam as the latter is transformed into a guided mode—assuming that the incoming gaussian is of the correct size.

From (3.74) and (3.75), it is known that the LP_{01} mode intensity varies with radius as $J_0^2(ur/a)$ inside the core, and as $K_0^2(wr/a)$ in the cladding. These two functions connect at $r = a$ to form the composite intensity profile as depicted in Fig. 3.13. This shape in fact closely fits a gaussian function having a width r_0, known as the *mode field radius*. This is defined as the radial distance from the core center to the $1/e^2$ point of the gaussian *intensity* profile. Note that this is the same distance that would be measured to the $1/e$ point of the gaussian electric field profile.

The gaussian mode approximation is useful because, apart from the fact that the intensity pattern is easier to visualize, the gaussian function is easier to use in constructing models of fiber performance. An example of this is the measure of mode confinement provided by the mode field radius. This enables one to establish with relative ease the requirements and tolerances for coupling between the fiber and a light source or between two fibers. The mode field radius is an important specification of a single-mode fiber and is often given (at a specified wavelength) instead of the core size. Under single-mode operation ($V < 2.405$), r_0 will always be larger than the core radius. Another example is in the construction of bending loss models using the gaussian mode field. Some accuracy is lost in these cases, however, since field strengths in the gaussian approximation are slightly lower at large distances into the cladding than is true for the actual LP_{01} mode, as is demonstrated in Fig. 3.13.

The approximation of LP_{01} by a gaussian involves the comparison of the LP_{01} mode field (presumed x-polarized) and a gaussian field having the same polarization [11]. The LP_{01} field from (3.72) is

$$
E_{1x} = \begin{cases} E_0 J_0(ur/a) \exp(-j\beta z) & r \leq a \\[2ex] E_0 \dfrac{J_0(u)}{K_0(w)} K_0(wr/a) \exp(-j\beta z) & r \geq a \end{cases} \tag{3.89}
$$

The gaussian field is expressed as

$$
E_{gx} = E_{g0} \exp\left(\frac{-r^2}{r_0^2}\right) \exp(-j\beta z) \tag{3.90}
$$

where r_0, the mode field radius, is to be determined. Both fields are assigned to carry the same power, P, evaluated through

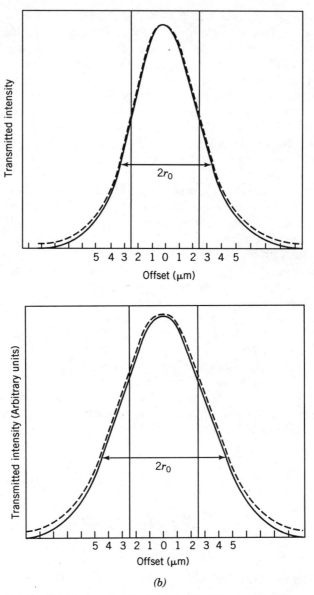

Figure 3.13. Gaussian fit to the intensity distribution of the LP_{01} mode (solid curve). The dotted curve is constructed from experimentally measured data at (a) $\lambda = 1.3\,\mu$m and (b) $\lambda = 1.55\,\mu$m. The solid vertical lines indicate the core boundaries. (Adapted from ref. 10.)

$$P = \int_0^{2\pi} \int_0^\infty \tfrac{1}{2} \operatorname{Re}\{E_{gx}H_{gy}^*\} r \, dr \, d\phi = \int_0^{2\pi} \int_0^\infty \tfrac{1}{2} \operatorname{Re}\{E_{1x}H_{1y}^*\} r \, dr \, d\phi$$

$$(3.91)$$

where $H_{gy} = E_{gx}/\eta_2$ and $H_{1y} = E_{1x}/\eta_2$, with $\eta_2 = \eta_0/n_2$.

Upon entering the fiber, the gaussian field will in general decompose into a series of guided and radiation modes, represented by

$$E_{gx} = \sum_{i=1}^\infty C_i E_{ix} \qquad (3.92)$$

where E_{ix} is the electric field of LP mode i (note that for the radiation modes, the summation becomes an integral). A similar expansion can be written for the associated magnetic fields, using the same coefficients, C_i. Each mode in these expansions carries some fraction of the overall power, P. By using (3.92) and the analogous expansion for H_{gx} in (3.91), the power is thus

$$P = \sum_{i=1}^\infty |C_i|^2 \int_0^{2\pi} \int_0^\infty \tfrac{1}{2} \operatorname{Re}\{E_{ix}H_{iy}^*\} r \, dr \, d\phi = \sum_{i=1}^\infty |C_i|^2 P \qquad (3.93)$$

where the orthogonality condition (3.86) has been used. It is thus observed that

$$\sum_{i=1}^\infty |C_i|^2 = 1 \qquad (3.94)$$

The fraction of the incident power that couples into LP_{01} is therefore given by $|C_1|^2$; this quantity (in the absence of reflective losses at the fiber input) is thus the coupling efficiency of a gaussian beam into a single-mode fiber.

LP_{01} is the mode that most closely resembles a gaussian, so one would expect C_1 to be the dominant coefficient with appropriate choice of r_0, whereas the contributions of the other modes in the series should be minimal. The procedure is thus to determine r_0 such that C_1 is maximized. This is accomplished by way of the following computation, in which orthogonality of the modes is again used:

$$\int_0^{2\pi} \int_0^\infty \tfrac{1}{2} \, \text{Re} \, \{E_{gx} H_{1y}^*\} r \, dr \, d\phi$$

$$= \int_0^{2\pi} \int_0^\infty \tfrac{1}{2} \, \text{Re} \left\{ \sum_i C_i E_{ix} H_{1y}^* \right\} r \, dr \, d\phi = C_1 P \qquad (3.95)$$

C_1 is thus found through

$$C_1 = \frac{\pi}{P} \int_0^\infty \text{Re} \, \{E_{gx} H_{1y}^*\} r \, dr \qquad (3.96)$$

Since the gaussian wave and the LP_{01} mode both carry power P, the field amplitudes appearing in (3.96) are both expressed in terms of P (see Problem 3.6) through

$$E_{g0} = \left(\frac{4\eta_0 P}{n_2 \pi r_0^2} \right)^{1/2} \qquad (3.97)$$

and

$$H_0 = \frac{E_0}{\eta_2} = \left(\frac{2n_2 P}{\pi \eta_0} \right)^{1/2} \frac{w}{a V J_1(u)} \qquad (3.98)$$

The net field expressions are obtained by multiplying the above amplitudes by the appropriate radial and z dependences given in (3.90) and (3.89), respectively. Substituting these results into (3.96) yields the final integral equation for the LP_{01} expansion coefficient:

$$C_1 = \frac{2\sqrt{2}}{r_0 a V} \left\{ \frac{w}{J_1(u)} \int_0^a J_0 \left(\frac{ur}{a} \right) \exp \left(-\frac{r^2}{r_0^2} \right) r \, dr \right.$$

$$\left. + \frac{u}{K_1(w)} \int_a^\infty K_0 \left(\frac{wr}{a} \right) \exp \left(-\frac{r^2}{r_0^2} \right) r \, dr \right\} \qquad (3.99)$$

For each choice of V in (3.99), r_0 is varied until the equation yields a maximum in C_1. The C_1 values obtained increase from 0.94 at $V = 1.2$, to nearly unity at $V = 2.4$, thus reflecting the accuracy of the gaussian approximation [11].

The results of the above process are closely approximated by the empirical formula [11]:

$$\frac{r_0}{a} \approx 0.65 + \frac{1.619}{V^{3/2}} + \frac{2.879}{V^6} \qquad (3.100)$$

This formula approximates the exact result to better than 1% accuracy. A simpler formula was obtained using a variational method [12], to be considered in Chapter 6:

$$\frac{r_0}{a} \approx \frac{1}{\sqrt{\ln V}} \qquad (3.101)$$

Both functions are shown in Fig. 3.14, in which the ratio of r_0 to core radius, a, is plotted as a function of V. As expected, r_0 is seen to decrease with increasing V, thus reflecting the principle of tighter mode confinement with frequency, as set forth in the previous section.

The mode field radius can be measured by any of a number of methods. One of these, the near field method, is shown in Fig. 3.15. In it, the gaussian output of a single-mode fiber is input to a lens, which in turn produces a magnified image of the surface of the fiber end. A detector with an illumination area that is

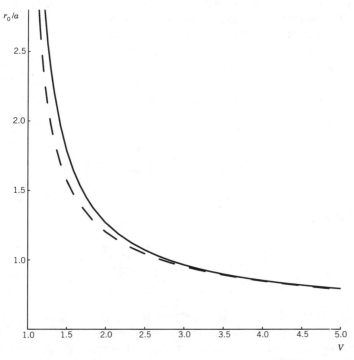

Figure 3.14. Normalized mode field radius, r_0/a, as a function of V for a step index fiber, as determined by the approximation formulas (3.100) (solid curve) and (3.101) (dashed curve).

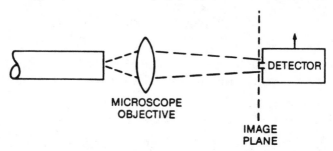

Figure 3.15. The near field method for measurement of the mode field radius [9].

much smaller than the image size is scanned across the image; the voltage from the detector is measured as a function of its position. The position at which the detector voltage falls to $1/e^2$ times the value at image center thus defines the mode field radius of the magnified image.

3.11. GAUSSIAN BEAM INPUT COUPLING

An application of the gaussian intensity distribution in a single-mode fiber is the problem of the input coupling of light from a well-collimated laser. The output intensity profile of such a laser that operates in the lowest-order transverse mode is in fact gaussian. To minimize coupling losses, the spot radius of the input beam should be matched to the fiber mode field radius. This is accomplished by using one or more lenses.

Assuming linear polarization (y direction) and z-directed propagation, the laser electric field takes the form

$$E_y = E_m \exp\left(\frac{-r^2}{r_g^2(z)}\right) \exp(-j\beta z) \qquad (3.102)$$

where the beam radius is

$$r_g^2(z) = r_0^2\left(1 + \frac{z^2}{z_0^2}\right) \qquad (3.103)$$

The z dependence of $r_g(z)$ is indicative of the beam divergence that occurs as it propagates away from the laser. The use of lenses will change the rate of divergence, or can cause the beam to converge to a focus at which the minimum spot radius, r_0, would occur. Figure 3.16 shows a gaussian beam that is focused between two lenses. The position $z = 0$ defines the location of the minimum

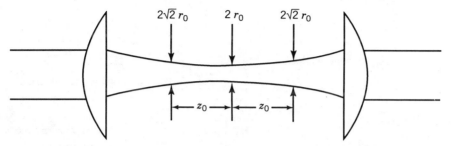

Figure 3.16. Focused gaussian beam, showing beam waist region within $z = \pm z_0$.

spot size; on either side of this position, the increase in spot radius is described by (3.103). The distance z_0, known as the *Rayleigh range*, defines the positions at which the beam size increases to $r_0\sqrt{2}$. The magnitude of z_0 depends on r_0 and the free-space wavelength through

$$z_0 = \frac{\pi r_0^2 n}{\lambda} \tag{3.104}$$

where n is the refractive index of the medium. In Fig. 3.16, the region defined by one Rayleigh distance on either side of $z = 0$ is known as the beam *waist*. A detailed treatment of gaussian beam theory can be found in a number of texts [13,14].

The strategy to optimize coupling is to focus the laser output to a minimum spot radius at the input face of the fiber. The spot radius is set equal to the fiber mode field radius. With the gaussian profiles of the input and guided fields matched in this way, optimum coupling is achieved. If the focused spot is too large, a fraction of the power would escape into the cladding; too small of a focus would produce a divergence angle within the fiber that is too large, which again would allow some of the power to escape through the cladding.

Figure 3.17 shows a simple coupling scheme in which a lens of focal length f has the effect of focusing the input beam to the required radius at the fiber input. The lens is positioned at the focus of the converging beam from the left. The focus will occur at a distance d to the right of the lens given by

$$d = \frac{f}{1 + (f/z_{02})^2} \tag{3.105}$$

where $z_{02} = \pi r_{02}^2 n/\lambda$ is the Rayleigh distance of the incoming beam. The ratio

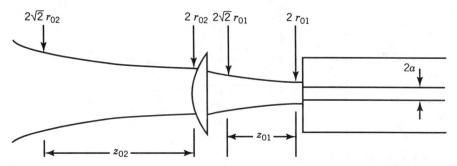

Figure 3.17. Scheme to focus a gaussian beam onto the end of a fiber, using a single lens.

of the spot radii at the lens focus and at its input is

$$\frac{r_{01}}{r_{02}} = \frac{f/z_{02}}{\sqrt{1 + (f/z_{02})^2}} \tag{3.106}$$

For the special case of an input beam that is collimated, $z_{02} \gg f$. Equation (3.106) thus becomes $r_{01} \approx f\lambda/\pi r_{02}n$, and (3.105) simplifies to $d \approx f$. In either case, the matching procedure involves first determining the required value of r_{01} from Fig. 3.14 or from (3.100), knowing V, and then choosing the lens focal length, based on the known value of r_{02}. Most often, very short focal lengths, on the order of 1 cm, are needed. This requires use of a good lens that is corrected for spherical aberration, a problem that becomes severe at short focal lengths. Common microscope objectives are usually adequate.

For coupling to a single mode fiber using an input beam radius that is not equal to the mode field radius, the fraction of the incident power that is coupled in (not including reflective losses) will be given by $|C_1|^2$. C_1 is found by evaluating (3.99) using the input gaussian beam radius for r_0.

PROBLEMS

3.1. Show, using propagation arguments for meridional rays, that the numerical aperture of a step index fiber in free space is given by Eq. (3.2).

3.2. Given the following modes—EH_{51}, TE_{04}, TE_{05}, HE_{17}, TM_{07}, EH_{37}, TM_{04}, HE_{71}, and HE_{27}—group the modes into degenerate sets and indicate the LP designation of each set. Also find the LP designation for any of the above modes that are not degenerate with others shown. Describe the intensity patterns for all LP modes that are found.

3.3. A certain dye laser is tunable over a wavelength range of 550 to 640 nm. The light is to be guided by a step index fiber in which $n_1 = 1.50$,

$n_2 = 1.48$, and $a = 0.9\,\mu m$. Determine which LP modes will propagate as the wavelength is varied over the given range. Specify the wavelength ranges for each mode. Over what range, if any, will the fiber operate single mode?

3.4. Under the weakly guiding approximation, the general eigenvalue equation for the propagating modes in a step index fiber was shown to be

$$\frac{J'_q(u)}{uJ_q(u)} + \frac{K'_q(w)}{wK_q(w)} = \pm q \left(\frac{1}{u^2} + \frac{1}{w^2} \right)$$

Using the above, along with appropriate recurrence relations for Bessel functions, derive the explicit eigenvalue equations as stated in Eqs. (3.41) through (3.43).

3.5. Show that the electric field of the LP_{01} mode is linearly polarized in either the x or y direction, depending on how the radial and angular components of the HE_{11} mode are superimposed. Refer to Eqs. (3.62) to (3.65), and thus derive (3.72) and (3.73). Also determine the magnetic field in the core for one electric field polarization in cartesian coordinates.

3.6. By appropriate integration of the Poynting vector over an infinite cross section in the xy plane, as exemplified in (3.83) and (3.84), perform the following tasks:

a) Show that the power carried by a gaussian beam propagating in a medium of index n_2, where the electric field is given by (3.90), is

$$P = \frac{n_2}{\eta_0} |E_{g0}|^2 \frac{\pi r_0^2}{4}$$

b) Show that the power carried by the LP_{01} mode in a step index fiber, with electric field given by (3.89), is

$$P = \frac{\pi a^2 n_2}{2\eta_0} |E_0|^2 J_1^2(u) \frac{V^2}{w^2}$$

where $n_1 \approx n_2$. The following integrals may be helpful:

$$\int x J_q^2(ax)\,dx = \frac{x^2}{2} [J_q^2(ax) - J_{q-1}(ax)J_{q+1}(ax)]$$

$$\int x K_q^2(bx)\, dx = \frac{x^2}{2}\, [K_q^2(bx) - K_{q-1}(bx)K_{q+1}(bx)]$$

3.7. Show for the LP_{01} mode that the ratio of the power carried in the core to the total power in the fiber is

$$1 - \nu = \frac{w^2}{V^2}\left(1 + \frac{J_0^2(u)}{J_1^2(u)}\right) \quad \text{for } LP_{01}$$

3.8. Show that as an alternative to the result of Problem 3.7, the power ratio can also be expressed as

$$1 - \nu = \frac{1}{2}\left(\frac{d(bV)}{dV} + b\right)$$

where $b = 1 - u^2/V^2$ is used. Begin by differentiating both sides of (3.48) with respect to V. This result is useful in interpreting dispersion behavior in single-mode fibers, as will be shown in Chapter 5.

3.9. An interesting interpretation of the normalized propagation constant, b, can be found by considering the following function:

$$F = \frac{\displaystyle\int_0^1 J_0(uR)R\, dR}{\displaystyle\int_0^1 J_0(uR)R\, dR + \int_1^\infty K_0(wR)J_0(u)/K_0(w)R\, dR}$$

This represents the integral of the LP_{01} mode field over the core cross section divided by the integral of the field over the entire fiber cross section, where $R \equiv r/a$. Using the integral forms $\int R J_0(ur)\, dR = (1/u)J_1(uR)$ and $\int R K_0(wr)\, dR = -(1/w)K_1(wR)$, along with (3.48), show that in fact $F = b$. Thus b provides a measure of the "field occupancy" within the fiber cross section.

3.10. Show, using Bessel function identities, that the eigenvalue equation for LP modes (3.45) can also be expressed in the form

$$u\frac{J_{l+1}(u)}{J_l(u)} = +w\frac{K_{l+1}(w)}{K_l(w)}$$

3.11.   Using the result of Problem 3.10, along with (3.45), show that Eq. (3.85) can be expressed in the form

$$\nu = \left(\frac{u^2}{V^2}\right)\left[1 + \left(\frac{w^2}{u^2}\right)\frac{J_l^2(u)}{J_{l+1}(u)J_{l-1}(u)}\right]$$

3.12.   Consider a fiber having core and cladding indices n_1 and n_2, operating at angular frequency ω_0. Using the graphical solution for TE, TM, and HE modes (Fig. 3.5), show the following:

a) The minimum value of β for the HE_{12} mode is $n_2 k_0$.

b) β can never be greater than $n_1 k_0$ for the HE_{12} mode.

3.13.   Two single-mode step index fibers are end-to-end coupled. One has core radius 3 μm and numerical aperture 0.1. The other has core radius 4 μm. Light of wavelength 1.3 μm propagates through the pair.

a) Determine the required numerical aperture of the second fiber, such that the coupling efficiency is optimized.

b) Will optimum coupling be lost or maintained as the wavelength is varied? Explain.

3.14.   A Nd:YAG laser (λ = 1.06 μm) produces a 1 cm diameter beam of gaussian intensity profile. It is desired to focus this beam into a single-mode fiber having indices n_1 = 1.46 and n_2 = 1.45 and core radius a = 2.0 μm. The focusing is to be done in two stages as shown in Fig. P3.1, where lens 1 is of 1.5 m focal length. The second lens is at the focal point of the first lens. Find the distances d_1 and d_2 and the focal length of lens 2 such that optimum coupling is achieved.

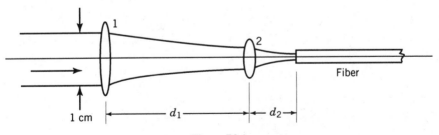

Figure P3.1

3.15.   The mode field radius of a certain step index fiber is specified to be 5.0 μm at wavelength λ = 1.3 μm. The cutoff wavelength of the fiber is given as λ_c = 1.2 μm. Determine the expected mode field radius at λ = 1.55 μm.

REFERENCES

1. T. Okoshi, *Optical Fibers*. Academic Press, New York, 1982.

2. E. Snitzer, "Cylindrical Dielectric Waveguide Modes," *Journal of the Optical Society of America*, vol. 51, pp. 491–498, 1961.

3. W. B. Jones, *Introduction to Optical Fiber Communication Systems*. Holt, Rinehart, and Winston, New York, 1988.

4. M. P. Lisitsa, L. I. Berezhinskii, and M. Ya. Valakh, *Fiber Optics*. Israel Program for Scientific Translations, New York, 1972.

5. D. Gloge, "Weakly Guiding Fibers," *Applied Optics*, vol. 10, pp. 2252–2258, 1971.

6. N. S. Kapany, *Fiber Optics, Principles and Applications*. Academic Press, New York, 1967.

7. R. A. Sammut, "Analysis of Approximations for the Mode Dispersion in Monomode Fibres," *Electronics Letters*, vol. 15, pp. 590–591, 1979.

8. H.-D. Rudolf and E.-G. Neumann, "Approximations for the Eigenvalues of the Fundamental Mode of a Step Index Glass Fiber Waveguide," *Nachrichtentechnische Zeitschrift (NTZ Communications Journal)*, vol. 29, pp. 328–329, 1976.

9. D. Franzen, "Tutorial on Single Mode Fiber Measurements," presented at the Optical Fiber Communications Conference, Atlanta, Georgia, 1986.

10. S.-J Jang, J. Sanchez, K. D. Pohl, and L. D. L'Esperance, "Fundamental Mode Size and Bend Sensitivity of Graded and Step-Index Single-Mode Fibers with Zero Dispersion Near 1.55 μm," *IEEE Journal of Lightwave Technology*, vol. LT-2, pp. 312–316, 1984.

11. D. Marcuse, "Loss Analysis of Single-Mode Fiber Splices," *Bell System Technical Journal*, vol. 56, pp. 703–718, 1977.

12. A. W. Snyder and R. A. Sammut, "Fundamental (HE_{11}) Modes of Graded Optical Fibers," *Journal of the Optical Society of America*, vol. 69, pp. 1663–1671, 1980.

13. D. Marcuse, *Light Transmission Optics*. Van Nostrand Reinhold, New York, 1982.

14. A. Yariv, *Quantum Electronics*, 3rd ed. John Wiley & Sons, New York, 1989.

4

Loss Mechanisms in Silica Fibers

Historically, the viability of fibers as long-range communication channels was contingent on reducing losses and dispersion to levels that would allow repeater separation distances on the order of kilometers. In silica-based fibers, it was found that, within certain regions in the 1.2 to 1.8 μm wavelength range, the simultaneous reduction of these two quantities could be achieved. The most extensive low-loss region, occurring over a continuum of wavelengths between 1.48 and 1.65 μm, represents a bandwidth of approximately 21,000 GHz—sufficient to support nearly 30 million voice channels.

A number of mechanisms that relate to a wide range of material and manufacturing parameters may contribute to the net loss in a fiber. These mechanisms are categorized as either *intrinsic* or *extrinsic*. The former arise from the fundamental material properties of the glasses used in fiber manufacture. Intrinsic losses are therefore closely related to the desired refractive indices and operating wavelengths; they can be reduced by appropriate choice of wavelength, and through various design compromises that involve different material compositions, as will be discussed. Extrinsic losses, on the other hand, arise from imperfections in the fabrication process that can essentially be eliminated with appropriate refinements. Sources of extrinsic loss include impurities that enter the glass during manufacture, causing additional absorption. Others relate to structural imperfections that can result in light scattering.

A more recent endeavor has concerned the development of fibers that amplify light at the transmission wavelength by way of gain induced in the fiber by additional "pump" light input at a shorter wavelength. Such fibers have been manufactured by doping the core material with rare-earth ions, such as neodymium, erbium, or praseodymium. In the case of erbium-doped fibers, the input of only 15 milliwatts (mW) of pump light at 1.48 μm was sufficient to provide enough gain for light at 1.55 μm to completely overcome the losses at that wavelength over a 10 km length. Significant amplification is also possible, making these fibers attractive for use as repeater sections in long-haul applications, such as undersea cable. Fiber amplifiers will be treated in Chapter 8.

Loss or gain is usually expressed as the ratio of input to output power levels per kilometer (km) length of fiber, expressed in decibels:

$$\alpha_p(\text{dB/km}) = \frac{10}{L} \log_{10} \left(\frac{P_{\text{in}}}{P_{\text{out}}} \right) \qquad (4.1)$$

where L is the fiber length. This is a convenient measure since loss and gain contributions from the various sources can be combined to yield the total value by simple addition. Current technology has enabled the complete elimination of extrinsic losses over a finite wavelength range beyond 1.4 μm. The fundamental intrinsic low-loss limit (determined principally by Rayleigh scattering) occurs at 1.55 μm; at this wavelength, α_p has a value of about 0.14 dB/km, corresponding to about a 3% loss over 1 km.

The importance of even minor reductions in fiber losses is understood by observing their effects on the performance of a transmission link. Consider a length L of fiber, having propagation loss α_p (in units of dB/km). The relation between the input and output (received) optical power in the fiber is specified through

$$P_r = P_{\text{in}} 10^{-0.1\alpha_p L} \qquad (4.2)$$

In most cases, the information is transmitted digitally in the form of a string of optical pulses that occur within equally spaced time slots. In the case of binary (on–off) modulation, the presence or absence of a pulse in a given slot corresponds to a "1" or a "0," with either of these two events occurring with equal probability. The content (or lack thereof) of each time slot thus corresponds to 1 bit of information. The data rate (or bit rate), B, is the number of time slots that are transmitted per second. The receiver at the fiber output requires a minimum energy, W_p, in each pulse in order to maintain a specified precision in interpreting ones and zeros. This precision is measured by the *bit-error rate* (BER), defined as the number of detection errors (or misinterpreted bits) that occur in a given time interval, divided by the number of bits transmitted in the same time interval. A link is typically designed to achieve a BER of 10^{-9}.

The above considerations lead to the required minimum *average* power at the receiver, $P_{r\,\text{min}}$, necessary to achieve the specified BER:

$$P_{r\,\text{min}} = \tfrac{1}{2} B W_p \qquad (4.3)$$

where the factor of $\tfrac{1}{2}$ accounts for the equal probability of the presence or absence of a pulse in each time slot. Suppose that at the fiber input end, the optical source is capable of generating P_{in} watts of coupled-in average power when modulated at *any* bit rate. The maximum allowable length of fiber, L_{max}, is then found from (4.2) by setting $P_r = P_{r\,\text{min}}$ and using (4.3):

$$L_{max} = \frac{10}{\alpha_p} \log_{10} \left(\frac{2P_{in}}{BW_p} \right) \qquad (4.4)$$

The above result shows that L_{max} is much more sensitive to changes in α_p than to changes in either P_{in} or W_p. Thus reducing α_p by a given factor will have a much greater effect on increasing L_{max} than would be obtained by similar improvements on either of the power levels. This simple principle has motivated the replacement of fiber systems operating at wavelengths in the vicinity of 1.3 μm with those that operate at 1.55 μm. In so doing, loss is reduced from about 0.5 dB/km to the neighborhood of 0.2 dB/km. A fiber link whose maximum length is determined by (4.4) alone is said to be *loss-limited*. The effects of dispersion, not included in this analysis, may reduce L_{max} in many cases. In addition, the wavelength transition from 1.3 to 1.55 μm introduces more dispersion, which will reduce to an extent the performance improvement as specified in (4.4). Dispersion effects and methods for their minimization will be presented in Chapters 5 and 6.

The present chapter explores the mechanisms by which loss can occur and how these relate to basic material properties and fabrication methods. Techniques for the computation and measurement of loss are presented. Finally, the problem of coupling loss is addressed, along with methods for its computation.

4.1. FABRICATION OF SILICA FIBERS

The basic challenges in manufacturing optical fibers involve achieving precise control of fiber dimensions and refractive index, while keeping losses at a minimum. Index control is achieved through changes in glass composition by introducing dopant materials into the silica host medium. This has been most successfully accomplished by vapor phase manufacturing techniques, which involve the oxidation of precursor gas mixtures to form material layers that build up the basic fiber structure, known as the preform. The fiber dimensions are established by stretching the molten preform into the completed fiber in the drawing process.

The development of vapor phase technology has resulted in four important process methods that are in use for fabricating silica fibers. These include modified chemical vapor deposition (MCVD), plasma-activated chemical vapor deposition (PCVD), outside vapor deposition (OVD), and vapor axial deposition (VAD). All of these use the same dopants to form the fiber structure. The MCVD process is described here to illustrate the basic chemistry. For a detailed review of all the above processes, the reader is referred to ref. 1.

The MCVD process begins with a hollow substrate tube of pure silica (SiO_2), which serves as part of the cladding in the completed fiber and also to contain the gases that take part in the glass-forming reactions. The process involves the deposition of a number of glass layers on the inside wall of the tube, to complete the formation of the cladding and to form the core. When deposition is complete, the tube is collapsed under vacuum to produce the finished preform. Figure 4.1 details the preform fabrication apparatus for MCVD. The silica tube rotates on a glass-working lathe while an oxyhydrogen torch is translated back and forth along its length to form a moving "hot zone" in which the chemical reactions occur. The input end of the tube is connected to the chemical delivery system, consisting of flow regulators and a mixing chamber for the various precursor gases. By-products of the reactions and nonreacted gases are collected at the output end.

Initially, layers of pure silica are deposited on the inside wall to complete the cladding. The added layers are generally of higher purity and are freer of defects and contaminants than the silica tube; the resulting fiber will thus have lower losses, provided most of the cladding power propagates in these deposited layers. This is in fact a requirement in low-loss single-mode fibers. Consequently, formation of the cladding accounts for most of the deposited material in the single-mode structures. The deposited cladding index is either the same as that of the substrate (for *matched cladding* fiber) or, in some cases, it is made less than that of the substrate by using appropriate dopants (forming *depressed cladding* fiber). The reasons for using the latter design will be discussed in Chapter 6. In addition, the silica cladding layers serve as a "water barrier," preventing the out-diffusion of loss-producing OH molecules into the reaction area from the tube. To form pure silica, the precursor gas $SiCl_4$ is used, which reacts with oxygen:

$$SiCl_4 + O_2 \rightarrow SiO_2 + 2Cl_2$$

The SiO_2 deposit initially consists of an unconsolidated layer of particles known as "soot." Further passes of the torch are necessary to sinter the soot layer, forming a homogeneous transparent film.

To form a material layer of higher refractive index, small amounts of germania (GeO_2) are introduced interstitially into the silica lattice. Over the wavelength ranges of interest, pure germania is of refractive index $n_{GeO_2} \approx 1.60$,

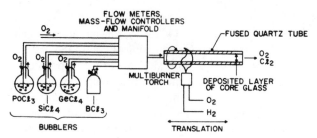

Figure 4.1. Preform manufacturing apparatus used in MCVD [2]. © 1980 IEEE.

whereas pure silica has index $n_{SiO_2} \approx 1.45$. Combinations of the two materials yield a net index between these two values that is determined with reasonable accuracy by the relation

$$n^2 = n_{SiO_2}^2 + C(n_{GeO_2}^2 - n_{SiO_2}^2) \qquad (4.5)$$

where C is the mole concentration of the GeO_2 dopant. To perform the doping process, an appropriate proportion of the precursor gas $GeCl_4$ is mixed with $SiCl_4$ in the reaction chamber. $GeCl_4$ oxidizes in the same manner as $SiCl_4$ to yield GeO_2 and Cl_2.

To produce layers of lower refractive index than SiO_2, fluorine (F) is used as a dopant. Again, the extent of index decrease depends on the mole fraction of F in the compound. A third dopant, P_2O_5 (introduced by way of the precursor gas $POCl_3$) is often used as a softening agent to reduce the viscosity of the silica–germania or silica–fluorine layers. This has the effect of increasing the uniformity of the dopant concentration in each layer and thus improves index control [3]. In addition, using P_2O_5 results in the deposition temperature being lowered by about 200°C, thus reducing the possibility of preform distortion that can occur if temperatures are too high. Another effect of P_2O_5 is to increase the refractive index slightly. Thus, to enable lower deposition temperatures in the cladding formation stage, matched cladding fibers are sometimes made using F and P_2O_5 dopants in such proportions that enable the refractive index changes introduced by each to exactly cancel.

Multiple layers are deposited until enough material accumulates to complete the desired fiber. To form the core region, this could require up to 70 layers for multimode fiber, and as few as only one layer for single-mode fiber. Indices can be altered from layer to layer to produce graded index profiles. On completion of the deposition phase, the tube is collapsed under high heat and vacuum to form the solid preform. The completed preform is positioned vertically in a drawing tower, where the fiber strand is pulled from the end of the preform under high localized temperature. Parameters such as the drawing rate and furnace temperature determine the fiber diameter; the latter is continually monitored, and the measured data are used to make real-time adjustments in the feed rate by means of a feedback control loop. Further down the tower, a molten polymer coating is applied to the cooled fiber; the coating is cured either in an additional furnace or with ultraviolet radiation. A typical drawing and coating machine is detailed in Fig. 4.2.

4.2. INTRINSIC LOSSES

There are three sources of intrinsic loss in pure silica that are important at visible and near-infrared wavelengths. These include the two resonances centered in the ultraviolet (UV) and mid-infrared and the mechanism of Rayleigh scattering.

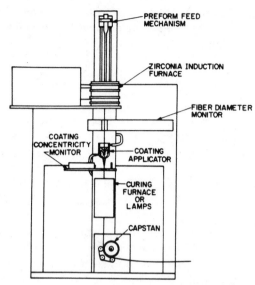

Figure 4.2. Schematic diagram of a typical fiber drawing and coating apparatus [4]. © 1980 IEEE.

All of these mechanisms are wavelength dependent, and their combined effects determine the basic wavelength range suitable for signal transmission.

The UV resonance is of electronic origin and is centered at $\lambda = 0.1\ \mu m$. The absorptive (imaginary) part of the susceptibility associated with this resonance is of sufficient width for the "tail" of the curve to yield appreciable absorption in the visible; but its effect is negligible in the near-infrared. The UV tail, or *Urbach edge* [5], tends to shift toward the infrared when dopant materials are added. For example, the addition of germania yields power loss (in dB/km) that is characterized through the empirical relation [6]

$$\alpha_{UV} = \left(\frac{1.542g}{46.6g + 60} \right) \exp \left(\frac{4.63}{\lambda} \right) \tag{4.6}$$

where g is the mole fraction of GeO_2 and λ is the wavelength (in μm). At typical doping levels ($g \approx 0.02$ for single mode), the addition of germania produces negligible loss increase at wavelengths of 1.3 μm and above.

Lattice vibrational modes of silica and dopant glasses are responsible for the second intrinsic loss mechanism. The vibrational modes produce absorptive resonances that are centered between 7 and 11 μm; the resonances for silica and germania occur at 9.0 and 11.0 μm, respectively. Significant broadening occurs as a result of anharmonic coupling between the numerous vibrational modes, resulting in an infrared (IR) absorption tail that extends into the transmission wavelength region in the near-infrared. As a rule, lighter atomic masses result in shorter resonant wavelengths; the effect is to shift the IR tail further into the transmission wavelength range. For this reason, light dopants such as B_2O_3 are avoided in fibers that are to operate at wavelengths between 1.2 and 1.8 μm.

Of the other dopants, GeO_2 has little influence on IR absorption but is likely to increase scattering losses. On the other hand, P_2O_5 may bond with OH (when the latter is present) to produce a stretch resonance with an overtone at 1.65 μm, thus resulting in substantial absorption at that wavelength. As a result, it is usually desirable to avoid the use of P_2O_5 in regions where the mode power is highest, unless OH can be reduced to extremely low levels. In single-mode fiber, it can still be advantageous to use when forming cladding layers that are situated at approximately $r = 2a$ or further from the fiber center.

In general, the power loss associated with IR absorption can be expressed as

$$\alpha_{IR} = A \exp(-a_{IR}/\lambda) \quad \text{dB/km} \tag{4.7}$$

where λ is expressed in μm. For $GeO_2{:}SiO_2$ glasses, values of A and a_{IR} that provide a reasonable fit to measured data are $A = 7.81 \times 10^{11}$ dB/km and $a_{IR} = 48.48 \mu$m [6].

The third intrinsic mechanism is Rayleigh scattering. This process is classically described by the excitation of and reradiation of light by atomic dipoles of dimensions that are much less than the incident wavelength. Consider a dipole oscillator of length dl, oriented along the z axis, where $dl \ll \lambda$ (Fig. 4.3); the dipole carries a time-harmonic current of amplitude $I = j\omega q$. With the dipole at the center of a spherical coordinate system, the far-zone ($R \gg \lambda$) electric and magnetic fields from the oscillator will be the angular components:

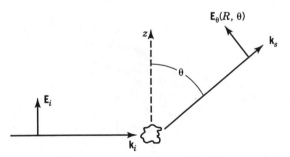

Figure 4.3. Schematic representation of dipolar scattering.

$$E_\theta = \frac{-pk_0^2 \sin \theta}{4\pi\epsilon_0 R} \exp(-jk_0R) = \eta_0 H_\phi \qquad (4.8)$$

where $p = q\,dl$ is the dipole moment magnitude. The above is in fact just the far-zone electric field from a Hertzian dipole, commonly used in the analysis of linear antennas. The radiated power density at a given value of θ, determined by the Poynting vector, is seen to be in proportion to the *fourth power* of the frequency. The scattering process consists of an incident wave exciting the dipole into oscillation, resulting in some absorption of the incident power. The dipole reradiates the power according to the pattern of (4.8). As frequency increases, more of the incident power is scattered as a result of the reradiation process. Such a scatterer, whose overall dimension is much less than the incident wavelength, is said to satisfy the *Rayleigh criterion*.

Pure silica glass is by nature amorphous, so that irregularities in the basic material structure can occur. The basic unit in the SiO_2 structure consists of a silicon atom surrounded by four oxygen atoms forming a tetrahedron. Each oxygen atom is shared by an adjacent tetrahedron; consequently, each silicon atom is associated with two complete oxygen atoms, as required by the chemical formula. Structural irregularities occur as a consequence of the large freedom in relative orientation that exists between two adjacent tetrahedra; an example of a resulting lattice arrangement is shown in Fig. 4.4. Such an arrangement is in dynamic thermal equilibrium at elevated temperatures (when the glass is in the "melt" stage); during manufacture the glass is rapidly cooled, leading to the high-temperature structure being "frozen in" to the final material. As a result, the material density exhibits random microscopic variations that correspond to those of the structure; local variations of refractive index occur with the changes in density, which act as scattering centers. Additional structural (and index) fluctuations arise from dopant molecules that are introduced into the SiO_2 lattice structure, leading to scattering losses that are dependent on dopant concentration. Index fluctuations arising from concentration or density

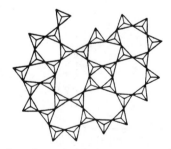

Figure 4.4. Cross-sectional view of a typical lattice structure of fused silica, showing possible connection arrangements for the SiO_4 tetrahedra [7].

fluctuations are of dimensions much less than a wavelength and thus satisfy the Rayleigh criterion.

Although detailed models of Rayleigh scattering processes in glass exist [8,9], scattering losses in a given fiber are typically determined through experimental measurement. The loss is expressed in decibels per kilometer through

$$\alpha_s = B/\lambda^4 \tag{4.9}$$

where B is the Rayleigh scattering coefficient, measured in dB/km $-$ μm^4. Since the core region of a fiber typically receives the dopants, scattering loss in general tends to increase with Δ. Figure 4.5 shows measured values of B as functions of Δ for some selected glass systems.

The net loss (in dB/km) due to intrinsic effects will be the sum of the contributions from each source; that is, $\alpha_I = \alpha_{UV} + \alpha_{IR} + \alpha_s$. At near-infrared wavelengths, α_{UV} is negligible compared to the other terms, so that

$$\alpha_T \approx A \exp(-a_{IR}/\lambda) + B/\lambda^4 \tag{4.10}$$

The minimum loss (and the wavelength at which it occurs) can be found by direct differentiation of (4.10) with respect to wavelength. Using the values

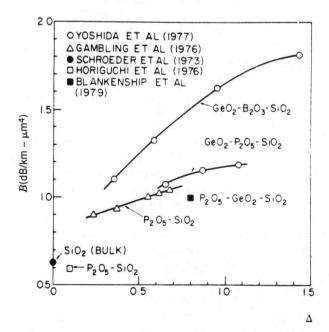

Figure 4.5. Plots of Rayleigh scattering coefficient, B, as functions of Δ for different glass compositions. All data were obtained from fiber samples, except for fused silica [10].

given above for A and B, it can be shown that the minimum loss wavelength is approximated by $\lambda_{min} \approx 0.03 a_{IR}$ μm [11]; the minimum loss corresponding to this wavelength is then determined principally by (4.9); that is, Rayleigh scattering is the dominant intrinsic loss mechanism in the vicinity of the lowest-loss wavelength.

4.3. EXTRINSIC LOSSES

Loss sources that are not associated with the fundamental material properties are categorized as extrinsic. These are generally associated with additional substances present in the glass compound that are not essential to the light-guiding properties of the fiber. They, and the associated losses, can in principle be removed with appropriate refinements in the fabrication process.

Cation impurities comprise one class of extrinsic loss. These include the transition metals, V, Cr, Ni, Mn, Cu, and Fe, all of which exhibit very strong absorption bands in the visible and near-infrared. In addition, the rare-earth impurities, Pr, Nd, Sm, Eu, Tb, and Dy, introduce absorptive losses that are important in the near-infrared. All must be reduced to concentration levels on the order of a few parts per billion (ppb) in order to reduce their loss contributions to negligible values. This has been achieved through the use of vapor phase processing technology.

The most difficult extrinsic loss contributor to remove is the hydroxyl group, OH, which enters the glass in the form of water vapor. The fundamental stretching resonance of OH is centered at wavelengths between 2.7 and 3.0 μm, the specific value depending on the glass composition. In glasses based on SiO_2, the center wavelength occurs at 2.75 μm. The OH vibrational mode is slightly anharmonic, which leads to oscillation at overtone frequencies. In silica, these occur at 1.38 μm (first harmonic), 1.24 μm, and 0.95 μm, thus illustrating the importance of the OH absorption at transmission wavelengths.

On either side of the 1.38 μm resonance, local loss minima occur at 1.3 μm and at 1.6 μm. An OH concentration of 1 part per million (ppm) by weight introduces loss of about 0.7 dB/km at 1.3 μm; on the other hand, the loss introduced by OH at 1.6 μm is less than 0.1 dB/km with the same concentration. With OH levels on the order of 0.5 ppm, low net losses between 0.2 and 0.3 dB/km can still be achieved at 1.55 μm. This demonstrates an additional advantage of operation at 1.55 μm over 1.3 μm, in that less stringent control of the OH content is needed with the former wavelength. The 1.38 μm OH band in fact determines the short-wavelength edge of the fiber transmission window at about 1.48 μm. Dopants generally tend to broaden the OH absorption peaks. As mentioned, use of P_2O_5 is particularly detrimental with OH present, since it introduces an additional absorption band at 1.65 μm. Thus very high control of OH content is needed in P-doped fibers.

Figure 4.6 shows a plot of the spectral attenuation in a GeO_2:SiO_2 fiber having an OH concentration of approximately 0.5 ppm. The effects of the intrinsic

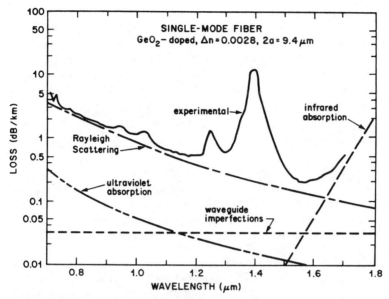

Figure 4.6. Spectral attenuation of a GeO_2–SiO_2 fiber showing the effects of intrinsic and extrinsic losses. The peak at 1.03 μm is the effect of bending loss of the LP_{11} mode near its cutoff wavelength (from ref. 13, adapted from ref. 12) © 1980 IEEE.

mechanisms considered above, in addition to OH absorption, are shown; the low-loss window centered at 1.55 μm is also apparent. The GeO_2:SiO_2 glass system is preferred over the use of other dopants since B_2O_3 tends to increase the infrared absorption, while P_2O_5 requires the reduction of OH levels typically by an order of magnitude. In practice, a compromise is often made, since the addition of P_2O_5 along with GeO_2 provides easier control of the refractive index. In such cases, the OH levels must be reduced to the 10 to 20 ppb range.

4.4. MACROBENDING LOSS

In Chapter 2, the concept of radiation loss in a bent slab waveguide was introduced, and it was shown that lossless propagation of light through a curved dielectric guide is generally not possible. The exception is the case of skew ray propagation in a round fiber, where, even though the ray reflects from a curved interface, no radiation loss occurs as long as the z component of the ray wavevector (β) is greater than or equal to $n_2 k_0$. If the fiber itself is bent, however, loss will occur to an extent that is inversely proportional to the bend radius.

There are in fact two types of bending loss that are associated with two different mechanisms. The first of these, known as *macrobending*, is associated with axial bends of relatively large radius (on the order of 1 mm or larger) in which the mechanism is the one discussed in Chapter 2. The second mechanism, known as *microbending*, involves the cumulative loss arising from small-magnitude "ripples" in an otherwise straight fiber; these are formed through small displacements of the fiber in directions that are transverse to its axis, and which have magnitudes on the order of micrometers. This effect can occur when the fiber comes into contact with a rough surface, such as that of a spool of inferior quality, or a rough plastic jacket material that would be used, for example, to bind several fibers together in a cable. Microbending is reduced significantly by coating individual fibers with a plastic material that conforms to surface irregularities.

Macrobending can be analyzed by expressing the cladding fields of a bent fiber as a superposition of those in the cladding of a bent slab waveguide [14]. For the latter structure, the geometry depicted in Fig. 2.8 is used, where the wave propagates in the direction of $\hat{a}_\phi$ and no z variation is assumed. The wave equation for the z-directed (transverse) electric field in the cladding becomes

$$\nabla^2 E_{z2} + n_2^2 k_0^2 E_{z2} = 0 \tag{4.11}$$

where the Laplacian is taken with respect to r and ϕ only. The product solution method is used, with an assumed solution (for one mode) of the form $E_{z2} = R(r)\Phi(\phi)$. The result is

$$E_{z2}(r,\phi) = A H_\rho^{(2)}(n_2 k_0 r) \exp(-j\rho\phi) \tag{4.12}$$

where $H_\rho^{(2)}(n_2 k_0 r)$ is the Hankel function of the second kind, and of order ρ, defined through

$$H_\rho^{(2)}(n_2 k_0 r) \equiv J_\rho(n_2 k_0 r) - j N_\rho(n_2 k_0 r) \tag{4.13}$$

It is observed from (4.12) that E_{z2} is oscillatory in r but exhibits exponential-like decay with radius at small values of r; the latter behavior arises from the presence of N_ρ, which approaches $-\infty$ as r approaches zero. The solution thus exhibits the features of an evanescent wave near the slab boundary but becomes oscillatory (radiative) at large values of r. The radiative behavior is readily seen by writing (4.12) for the case of large radius, using the large argument approximation for the Hankel function:

$$E_{z2}(r,\phi) \approx A \sqrt{\frac{2}{\pi n_2 k_0 r}} \exp[j(1 + 2\rho)\pi/4] \exp(-j n_2 k_0 r)$$
$$\cdot \exp(-j\rho\phi) \qquad \text{large } r \tag{4.14}$$

The Hankel function of the first kind, $H^{(1)}$, is also a solution to (4.11), but it describes a wave that radiates inward, instead of outward, and so is not used in the solution.

The value of ρ is determined by requiring that planes of constant phase occur along radial planes, as depicted in Fig. 2.8. Phase shift along the guided wave propagation direction is thus determined by the change in ϕ only, regardless of radius. Letting β be the mode propagation constant evaluated at the slab center (radius r_m), it must follow that $\rho = \beta r_m$. Note that ρ need not be an integer, so that Hankel functions of noninteger order will be found in the field solutions.

The extension of the slab waveguide results to the fiber case involves expressing the cladding mode field of the bent fiber as a superposition of solutions of the form of (4.12) [14]. The geometry is indicated in Fig. 4.7, where the fiber bend radius is r_m. The bent fiber axis lies in the xy plane, with the z axis at the center. A mathematical surface in the form of a cylindrical shell, centered on the z axis, is positioned with its side tangent to the outermost surface of the fiber core. Within the cylinder, the fiber field solutions are approximated as those of the straight fiber; outside the cylinder, the field is expressed as a superposition of bent slab waveguide solutions.

The coefficients of the terms in the expansion are evaluated by equating the two solution forms at the cylinder boundary. With the outside field determined, the power lost through radiation from the bend is found by evaluating the Poynting vector in the outside region, and then integrating it over a cylindrical surface at large radius, which is of infinite length in $\pm z$. The radiated power, P_{rad}, can be expressed in terms of the input power to the fiber, P_{in}, and a power loss coefficient, α_b, through

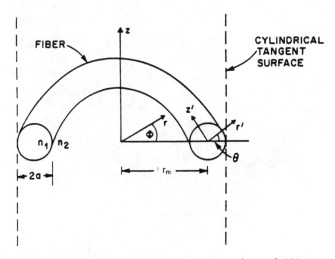

Figure 4.7. Bent fiber geometry. (Adapted from ref. 14.)

$$P_{rad} = P_{in}[1 - \exp(-\alpha_b z')] \tag{4.15}$$

where z' is measured along the bent fiber axis. In carrying out the above procedure, the power loss coefficient is found and can be expressed for a given LP mode in terms of the familiar parameters defined in Chapter 3 [14]:

$$\alpha_b = \frac{\sqrt{\dfrac{\pi}{a}}\, u^2 \exp\left[-\dfrac{2}{3}\left(\dfrac{w^3}{a^3\beta^2}\right) r_m\right]}{e_l w^{3/2} V^2 \sqrt{r_m} K_{l-1}(w) K_{l+1}(w)} \qquad e_l = \begin{cases} 2 & (l = 0) \\ 1 & (l \neq 0) \end{cases} \tag{4.16}$$

A plot of (4.16) is shown in Fig. 4.8 for the LP_{01} mode ($l = 0$) in a fiber having typical structural parameter values. Loss is seen to increase as wavelength increases and as bend radius decreases. Both of these features are intuitively reasonable: in the former case, a fiber mode becomes less confined to the core as wavelength increases and would thus be more susceptible to bending loss; in the latter case, a tighter bend results in the critical radius (at which a radial component of $\mathbf{k}$ arises) occurring at a smaller value of r. It was shown in Chapter 2 that a smaller critical radius results in a greater percentage of power transmitted

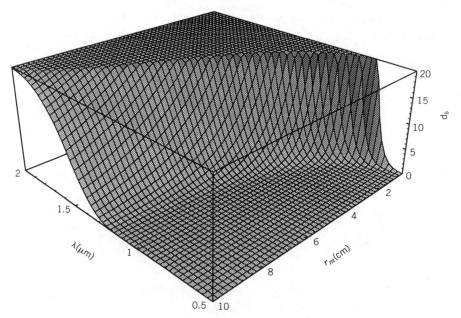

Figure 4.8. Macrobending loss (in dB/cm) for LP_{01}, calculated from (4.16) for a fiber of core radius $a = 2\,\mu\text{m}$, core index $n_1 = 1.454$, and cladding index $n_2 = 1.450$.

in a plane wave that is incident on a curved interface. Hence more power is lost from inside the core as the bend radius decreases.

Bending loss is also mode dependent, since for a given wavelength and bend radius, different modes will have different degrees of confinement within the fiber core. Figure 4.9 shows loss as a function of bend radius as calculated from (4.16) for the LP_{01} and LP_{11} modes in the same fiber that was used in the computation of Fig. 4.8. The more loosely confined LP_{11} mode is more susceptible to bend loss; consequently, over a certain range of bend radii, LP_{01} can experience very little loss, whereas LP_{11} is almost completely radiated away. A bend thus provides a simple way of filtering out most of any unwanted LP_{11} power in an otherwise bimoded fiber.

It has been assumed throughout that the mode field in the bent fiber (existing within the cylinder of Fig. 4.7) is not perturbed by the bend—the field solution for the straight fiber is assumed. Actually, there will be mode distortion with bending that becomes significant for small bend radii. The effect is a "migration" of the mode field distribution into the cladding, in a way that is analogous to the effect of centrifugal force on an object that moves through a turn. For example, the LP_{01} mode in a bent fiber maintains its gaussian shape, but the peak of the gaussian moves away from the fiber axis toward the outside. Using a perturbational approach to solve the wave equation for a bent fiber [15], the peak of the gaussian is found to be displaced through a distance given by the simple formula [16]

$$d_s = \frac{V^2 r_0^4}{\Delta r_m a^3} \tag{4.17}$$

where r_0 is the mode field radius in the straight fiber. In addition, the mode

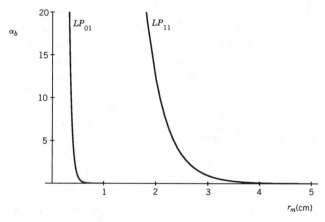

Figure 4.9. Macrobending loss (in dB/cm) calculated from (4.16) for the LP_{01} and LP_{11} modes in the fiber specified in Fig. 4.8.

field radius increases as the bend radius decreases according to [16,17]

$$\delta r_0 = \frac{(r_0' - r_0)}{r_0} = \frac{1}{2} \left(\frac{d_s}{r_0} \right)^2 \tag{4.18}$$

where r_0' is the corrected mode field radius as a result of bending. As a result, the fraction of power that is carried in the cladding increases, which increases the loss.

Another assumption made in deriving (4.16) is that the cladding is of infinite radius. In practice, the finite cladding radius leads to the coupling of the radiated power into "whispering gallery" modes in the bent cladding [18]. These in turn can feed power back into the guided modes, thus reducing the overall loss.

In view of the above, (4.16) is of value in that it demonstrates the behavior of bending loss in a useful way, but it nevertheless yields semiquantitative results. The information learned is that, in general, bends of radii on the order of a few centimeters are undesirable. Furthermore, bend losses are minimized if the mode is tightly confined. For this reason, V and Δ should be made as large as possible within a reasonable safety margin to assure single-mode operation. Typical V values in single-mode fibers range from about 1.8 to 2.2.

4.5. MICROBENDING LOSS

Microbending is in fact one case of a more general loss mechanism associated with *microdeformations* [19]. These could arise either in the form of small axial displacements of the type described above, or they could occur as fluctuations in the core radius around a straight axis—or some combination of the effects. In any case, the deformations occur in a random fashion along the z axis. The rate at which the fluctuations occur with z, however, is limited to a relatively narrow band of spatial frequencies. The resulting *space-harmonic* structure of the fiber deformations provides a mechanism by which the power in a guided mode can couple into an adjacent mode or into radiation modes. In the latter case, the guided mode power is lost to an extent that depends on the strength of the coupling.

Consider two guided modes having propagation constants β_1 and β_2. Power in one mode will couple into the other by way of the periodic deformations when the following condition is satisfied:

$$|\beta_2 - \beta_1| \approx \theta_0 \tag{4.19}$$

where θ_0 is the center spatial frequency of the deformations in radians per meter. The effect can be thought of in terms of grating diffraction concepts. Specifically, a grating having spatial frequency $\theta_0/2\pi$ will diffract an incident plane wave, yielding an outgoing wave that propagates at a different angle. The incident and diffracted waves are characterized by the components of their propagation constants *along the plane of the grating*, β_1 and β_2. Equation (4.19) is in fact a form of the grating equation. In the case of the fiber, a given mode, identified by a certain ray angle, will "diffract" into an adjacent mode by way of the deformation "grating." Typically, the deformations occur at spatial frequencies that are only high enough to enable significant coupling between adjacent modes and between the highest order mode and the radiation modes. Since the lowest order radiation mode will have propagation constant $n_2 k_0$, the condition for microdeformation loss to occur in a fiber is found from (4.19) to be

$$\beta - n_2 k_0 \approx \theta_0 \tag{4.20}$$

where β is the mode propagation constant.

Since the deformation function, $f(z)$, is a random process, its explicit form in a given fiber cannot be known. Instead, the behavior is modeled using the correlation function, $R(\delta) = \overline{f(z)f(z+\delta)}$, where the overline indicates that an ensemble average (over many values of z) is taken. Since fibers are manufactured using fairly tight process control over long time periods, it is intuitively clear that $f(z)$ will be *stationary*; that is, $R(\delta)$ will depend only on δ and not on z.

Various models of $R(\delta)$ have been used, among which are the exponential and gaussian functions:

$$R(\delta) = \sigma^2 \exp\left[-|\delta|/L_c\right] \tag{4.21}$$

$$R(\delta) = \sigma^2 \exp\left[-(\delta/L_c)^2\right] \tag{4.22}$$

where σ is the root mean-square (rms) deviation of $f(z)$, and L_c is a correlation length parameter. The Fourier transforms of (4.21) and (4.22) give the associated spectral density functions; for example, the transform of (4.18) (upon translation by θ_0) yields the Lorentzian function

$$F^2(\theta) = \frac{2\sigma^2}{L_c[\Delta\theta^2 + 1/L_c^2]} \tag{4.23}$$

where $\Delta\theta = \theta - \theta_0$. The deformations are thus characterized as exhibiting a range of spatial frequencies, $\Delta\theta = 1/L_c$, about the center frequency θ_0.

Better flexibility in comparing theory to experiment has been achieved by using a more generalized form of (4.23), in the form of a power law [20]:

$$F^2(\theta) = \frac{mL_c\sigma^2 \sin(\pi/m)}{1 + (L_c\Delta\theta)^m} \qquad (4.24)$$

which reduces to (4.23) when $m = 2$. Typically, values of m between 4 and 5 have yielded the most favorable agreement with the performance of actual fibers [21]. Although it is not possible to predict the exact form of $F^2(\theta)$, its approximations such as (4.24) have found use in the interpretation of results and for providing means to compare different fiber and cable designs.

The power loss through microbending is found by determining the power that is coupled from a given guided mode into all the radiation modes. This is done by multiplying $F^2(\theta)$ by the coupling coefficient [22] that links the guided mode to each radiation mode, and then summing the results. Since the radiation modes form a continuum of propagation constant values within the range $-n_2k_0 < \beta < n_2k_0$, the summation becomes an integral. The result is shown below for any LP mode having azimuthal mode number l [20]:

$$\alpha_d = \frac{\Delta w_l^2 J_l^2(u_l)a^2}{\pi^2 |J_{l-1}(u_l)J_{l+1}(u_l)|} \int_{-n_2k_0}^{n_2k_0} |\beta|F^2(\beta_l - \beta)$$

$$\times \left\{ \frac{J_{l-1}^2(u)}{|uJ_l(u)H_{l-1}^{(1)}(w) - wJ_{l-1}(u)H_l^{(1)}(w)|^2} \right.$$

$$+ \left. \frac{J_{l+1}^2(u)}{|uJ_l(u)H_{l+1}^{(1)}(w) - wJ_{l+1}(u)H_l^{(1)}(w)|^2} \right\} d\beta \qquad (4.25)$$

where u and w are expressed in the form given by (3.14) and (3.15) in terms of the integration variable, β; u_l and w_l are similarly expressed in terms of the fixed propagation constant for the mode, β_l.

Various approximate forms of (4.25) exist for special conditions. Of particular interest is the behavior of the LP_{01} mode under single-mode operation. Plots of (4.25) for this case are shown in Fig. 4.10 [20]. The loss values shown are normalized to the quantity $k_0^4 F^2(\beta_{01} - n_2k_0)$, and the curves are plotted as functions of V for a number of m values. An empirical fit to the curves is given by the formula [20]

$$\alpha_d \approx \frac{n_2 k_0^4 F^2(\beta_{01} - n_2k_0)}{(m-1)(m-2)} \left(\frac{\beta_{01} - n_2k_0}{k_0} \right)^3$$

$$\cdot \left(V + 7.58 \times 10^{-5} \frac{m^9 V^m}{e^{1.7m}} \right) \qquad (4.26)$$

The above is valid over the range $m \geq 4$.

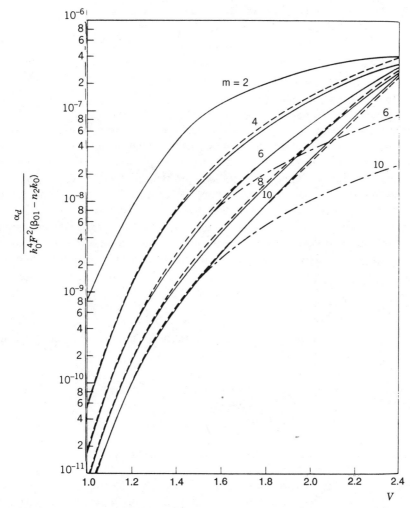

Figure 4.10. Normalized microbending loss as calculated from (4.25) (solid curves) and from (4.26) (dashed curves) for the single-mode fiber case [20].

The wavelength and mode size dependences of microbending loss in single-mode fibers are of significant interest. Although the trend of the curves in Fig. 4.10 is generally upward for increasing V, the actual loss (upon removing the normalization to k_0^4) is found in fact to increase with *increasing* wavelength. This is reasonable, since the mode becomes less confined as λ increases and so is likely to encounter a greater number of imperfections. Petermann [17,23] developed a model based on reasoning similar to that presented above, but which provides the explicit wavelength and mode field radius dependence on microbending loss in single-mode fibers. This result is

$$\alpha_d \; \propto \; \frac{r_0^{2(1+2p)}}{\lambda^{2(1+p)}} \tag{4.27}$$

where p is a shape parameter for the spectral density function used in the analysis [17]. The value of p lies between 1 and 2 for most realistic cases, but the best accuracy of (4.27) occurs for the $p = 1$ case. In any event, the model demonstrates the need to maintain a relatively small mode field radius to keep bending loss to a minimum. A trade-off exists, in that coupling loss will increase as r_0 decreases, leading to the need for some compromise.

Multimode fibers generally experience lower microbending loss, which is essentially independent of wavelength. Loss is lower because only the highest order modes (which carry a small fraction of the power) are closely coupled to the radiation modes; the wavelength independence arises from the relative insensitivity of the power confinement with wavelength in the multimode case.

4.6. COUPLING LOSS

Coupling light into a fiber requires consideration of the source emission characteristics and the design of the fiber. Many factors determine the choice of fiber and source for a given system or experiment; the coupling issue is just one of these. Therefore coupling efficiency that is less than ideal must usually be tolerated. The efficiency is defined as

$$\eta \; = \; \frac{P_c}{P_s} \tag{4.28}$$

where P_s is the total power emitted by the source, and P_c is the power that is coupled into the fiber. The power loss from coupling in decibels is given in terms of these values as

$$\alpha_c \; = \; 10 \log_{10} \left(\frac{P_s}{P_c} \right) \tag{4.29}$$

Among the available light sources, there is significant variation in emission characteristics. The most desirable source is a well-collimated laser, whose beam profile is easily tailored to produce optimum coupling.

Semiconductor lasers, while being attractive because of their small size and low operating cost, often emit noncollimated and noncircular beams that require more sophisticated means to produce good coupling. The least expensive but most undesirable source in terms of coupling difficulties is a light-emitting diode (LED). Its broad angular range of emission and large surface area provide substantially lower coupling efficiencies when compared with laser sources—particularly for single-mode fibers. Nevertheless, LEDs are usually the most favorable source for low-budget, low-bandwidth systems.

In determining the optimum beam shape for coupling, it is best to first observe how the fiber *emits* light. The intensity profile and maximum angle of emission from the fiber end is precisely the configuration of light that should be input to the fiber to be sure that the best coupling efficiency is achieved. This principle is perhaps best exemplified in the problem of coupling collimated laser light into a single-mode fiber, as was described in Chapter 3. In that case, the best results were found to occur when the input gaussian beam spot size matched that of the fiber mode. Deviations from this condition lead to a reduction of the LP_{01} expansion coefficient, C_1 (as calculated in Eq. (3.96)). The quantity $|C_1|^2$ is in fact the transmission coefficient, or coupling efficiency, assuming no reflective losses from the fiber input face. For nongaussian sources, a similar computation can be performed, leading to values of C_1 for these cases that will in general be lower. This is because there is less of a match between the field distribution of the input light and that of the fiber.

In the multimode fiber case, the field distribution is no longer gaussian, but the fiber power can be assumed to exist almost entirely within the core. Consequently, it can be assumed that incident light will be coupled in with unity efficiency (ignoring reflective losses) if it impinges on the fiber end face at $r < a$, and if it propagates at an angle to the z axis that is less than the fiber numerical aperture angle, $\theta_{N.A.}$. Coupling efficiency may suffer because (1) a given source may emit over an angular range that includes values greater than $\theta_{N.A.}$, and (2) the emitting area of the source may be larger than the fiber core area. Losses arising from the first effect are minimized by positioning the fiber end as close to the source as possible. The use of a lens between source and fiber will enable some separation without compromising the efficiency but will not reduce losses from the second effect.

Figure 4.11 shows the surface of a light source that is near to the cleaved end

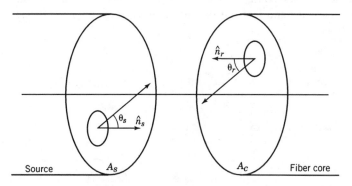

Figure 4.11. Geometry for end-to-end coupling between source and fiber.

of a fiber core. The source area is subdivided into differential areas, dS_s, each of which emits light over an angular range, $0 < \theta_s < \pi/2$. The receiving surface of the fiber will accept light over the range of angles, $0 < \theta_r < \theta_{N.A.}$. The directionality of the source emission is characterized by a *photometric brightness* function, which indicates the radiated power in a given direction, per source area, per solid angle subtended at the observation point. For example, the surface of a LED can be thought of as a uniform distribution of point sources, each of which radiates isotropically. The brightness observed from the diode surface will thus be only a function of the effective surface area that is seen from the observation point; an observer facing the diode directly would see the maximum brightness, whereas if it is viewed edge-on, the observer would see nothing. The observed area cross section decreases with angle to the normal as $\cos \theta_s$. The brightness function for the diode is thus written

$$B = B_0 \cos \theta_s \quad \text{W/cm}^2 - \text{sr} \tag{4.30}$$

where B_0 is a constant. The differential power received at a location on the fiber surface from differential area dS_s will be

$$dP_s = B_0 \cos \theta_s \, dS_s \, d\Omega \tag{4.31}$$

The differential solid angle is given by $d\Omega = \sin \theta_s \, d\theta_s \, d\phi$, where ϕ is the spherical coordinate azimuthal angle measured in a plane parallel to the emitter or fiber surfaces. The total power emitted into the fiber from differential area dS_s is

$$dP_c = \int_0^{2\pi} \int_0^{\theta_{N.A.}} B_0 \cos \theta_s \sin \theta_s \, d\theta_s \, d\phi \tag{4.32}$$

The total power coupled into the fiber is now found by integrating (4.32) over the surface area of the source or the fiber core, *whichever is less*. Assuming the fiber core is smaller and that the brightness is uniform over the diode surface, (4.32) becomes

$$P_c = 2\pi A_c \int_0^{\theta_{N.A.}} B_0 \cos \theta_s \sin \theta_s \, d\theta_s = \pi A_c B_0 \sin^2 \theta_{N.A.} \tag{4.33}$$

The total power radiated by the source is

$$P_s = 2\pi A_s \int_0^{\pi/2} B_0 \cos \theta_s \sin \theta_s \, d\theta_s = \pi A_s B_0 \tag{4.34}$$

The coupling efficiency is now

$$\eta = \frac{P_c}{P_s} = \frac{A_c}{A_s} \sin^2 \theta_{N.A.} \qquad (4.35)$$

One way to improve the efficiency is to use a more directional source, which radiates a greater percentage of its power within the numerical aperture angle of the fiber. The increase in directionality can often be modeled by a brightness function which includes a power of $\cos \theta_s$; for example, $B = B_0 \cos^n \theta_s$, where n is adjusted to approximately fit the observed radiation pattern.

As previously stated, a lens can be used to increase the separation between source and fiber or to enable more light to be gathered from the source. A simple arrangement is shown in Fig. 4.12, in which the lens diameter is assumed much larger than either the source or fiber core diameter. To optimize coupling using a lens, it is desired to *image* the source area onto the fiber core area, such that the image and core have equal areas. The ratio of image and source diameters is the *magnification* of the lens:

$$M \equiv \frac{d_i}{d_s} = \frac{L_2}{L_1} = \frac{\tan \theta_l}{\tan \theta_i} \qquad (4.36)$$

The distances L_1 and L_2 are related to the lens focal length through

$$\frac{1}{L_1} + \frac{1}{L_2} = \frac{1}{f} \qquad (4.37)$$

The lens must be positioned as close as possible to the source, so that it can receive the most light. In this way, θ_l is made as large as possible. With a given magnification requirement (determined by the core and source diameters), any decrease in L_1 must be accompanied by a proportional decrease in L_2, to

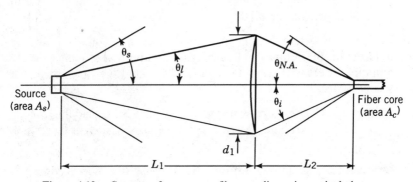

Figure 4.12. Geometry for source-to-fiber coupling using a single lens.

maintain constant M. The minimum allowable L_2 is determined by the maximum θ_i, which in turn can be no greater than the numerical aperture angle, $\theta_{N.A.}$. Thus the minimum source-to-lens distance will be $L_{1\,min} = L_{2\,min}/M = d_l/(2M \tan \theta_{N.A.})$, where d_l is the lens diameter. Under this condition, the maximum lens angle will be given by $\theta_{l\,max} = \tan^{-1}(d_l/2L_{1\,min}) = \tan^{-1}(M \tan \theta_{N.A.})$, which is independent of the lens diameter.

Ideally, light from the source that is gathered by the lens should be completely coupled into the fiber, provided the above requirements on distances are met. The coupling efficiency is thus determined by taking the ratio of the source power intercepted by the lens to the total radiated power. Assuming a LED source, the former is given by

$$ P_c = 2\pi A_s \int_0^{\theta_l} B_0 \cos \theta_s \sin \theta_s \, d\theta_s \qquad (4.38) $$

It is interesting to compare the efficiencies achieved for the cases of end-to-end coupling between source and fiber, and when an intervening lens is used. It can be shown (Problem 4.7) that for a general brightness function of the form $B = B_0 \cos^n \theta_s$, no improvement in efficiency can be achieved by using a lens if the source area is greater than the fiber core area. If the reverse is true, then use of a lens in the manner described above will produce an increase in efficiency by approximately a factor of M^2. This of course does not include reflection losses due to the lens, and it is also assumed that the lens is aberration-free.

PROBLEMS

4.1. An important figure of merit for a transmission link is the product of the bit-rate and the length, $B \times L$. Consider the bit-rate $\times$ distance product as applied to a loss-limited link, as described by (4.4). Considering (4.4) only, what effects (if any) will limit $B \times L$, assuming that P_{in} and W_p are fixed? In reality, what are the factors (if any) that limit the maximum value of B that can be used?

4.2. Using (4.10), determine an expression for the minimum loss wavelength in terms of the infrared and Rayleigh scattering loss parameters. Determine the effect of λ_{min} of (a) increased infrared loss and (b) increased scattering loss.

4.3. A silica-based fiber is constructed with a pure SiO_2 cladding and a 2% GeO_2-doped silica core. Determine (a) the core refractive index and (b) the loss at $\lambda = 1\,\mu m$ from UV absorption.

4.4. Estimate the loss (in dB/km) for a $GeO_2{:}P_2O_5{:}SiO_2$ fiber having $\Delta = 0.6$ at wavelength $\lambda = 1.3\,\mu m$.

4.5. In the single-bend attenuation technique for measuring the cutoff wavelength (Section 3.9), explain the reduction in loss with decreasing wavelength observed in the lower wavelength region of Fig. 3.12b.

4.6. Fiber loss is often measured by means of the *cutback method*. In it, the output power of a length of fiber is first measured. Then a section of length L is cut from the output end. The output power is then measured from the end of the shortened fiber, with the original source and launch optics undisturbed. Determine an expression for the power loss (in dB/km) for the fiber in terms of the two measured power values and the length of the removed fiber section, L.

4.7. Consider a light source of surface area A_s, characterized by photometric brightness function $B = B_0 \cos^n \theta_s$, where n is an integer. The source is to be coupled to a step index optical fiber of core area A_c, using two methods: (1) the fiber is to be placed end-to-end with the source, and (2) a lens is to be used between the source and fiber. Show that no improvement can be achieved in coupling efficiency by using a lens if $A_s > A_c$. Also show that when $A_c > A_s$, use of a lens will improve the efficiency over the end-to-end coupled case by approximately a factor of M^2, where M is the lens magnification. Refer to Fig. 4.12 and assume that all angles are small, such that trigonometric functions can be approximated by at most two terms in a series expansion.

4.8. Two multimode fibers are end-to-end coupled as shown in Fig. P4.1. Fiber 1 has core area A_1 and numerical aperture $N.A._1$. The Fiber 2 parameters are A_2 and $N.A._2$, where $A_2 < A_1$, and $N.A._2 > N.A._1$. Both numerical apertures are small, such that $\sin \theta_{N.A.} \approx \theta_{N.A.}$. The emission functions of the fiber end surfaces are modeled as Lambertian, with brightness $B = B_0 \cos \theta$, valid over the range $\theta < \theta_{N.A.}$. Determine the coupling efficiency of the joint for light incident from (a) Fiber 1 and (b) Fiber 2, in terms of the given parameters. In principle, could a lens be used between fibers to improve the coupling efficiencies?

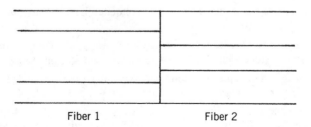

Fiber 1   Fiber 2

Figure P4.1

REFERENCES

1. S. R. Nagel, "Fiber Materials and Fabrication Methods," in *Optical Fiber Telecommunications II*, S. E. Miller and I. P. Kaminow, eds. Academic Press, San Diego, CA, 1988.

2. J.B. MacChesney, "Materials and Processes for Preform Fabrication—Modified Chemical Vapor Deposition and Plasma Chemical Vapor Deposition," *Proceedings of the IEEE*, vol. 68, pp. 1181–1184, 1980.

3. D. L. Wood, K. L. Walker, J. B. MacChesney, J. R. Simpson, and R. Csencsits, "Germanium Chemistry in the MCVD Process for Optical Fiber Fabrication," *IEEE Journal of Lightwave Technology*, vol. LT-5, pp. 277–285, 1987.

4. L. L. Blyler, Jr. and F. V. DiMarcello, "Fiber Drawing, Coating, and Jacketing," *Proceedings of the IEEE*, vol. 68, pp. 1194–1198, 1980.

5. F. Urbach, "The Long-Wavelength Edge of Photographic Sensitivity and of the Electronic Absorption of Solids," *Physical Review*, vol. 92, p. 1324, 1953.

6. S. R. Nagel, J. B. MacChesney, and K. L. Walker, "An Overview of the Modified Chemical Vapor Deposition (MCVD) Process and Performance," *IEEE Journal of Quantum Electronics*, vol. QE-18, no. 4, pp. 459–476, 1982.

7. P. K. Cheo, *Fiber Optics, Devices and Systems*. Prentice-Hall, Englewood Cliffs, NJ, 1985.

8. R. Olshansky, "Propagation in Glass Optical Waveguides," *Reviews of Modern Physics*, vol. 51, no. 2, pp. 341–367, 1979.

9. M. E. Lines, "Ultralow-Loss Glasses," *Annual Review of Material Science*, vol. 16, pp. 113–135, 1886.

10. R. Olshansky, "Optical Properties of Waveguide Materials for 1.2 and 1.8 μm," in *Physics of Fiber Optics—Advances in Ceramics Series, vol. 2*, B. Bendow and S. S. Mitra, eds. Plenum Press, New York, 1981, pp. 40–54.

11. M. E. Lines, "The Search for Very Low Loss Fiber-Optic Materials," *Science*, vol. 226, pp. 663–668, 1984.

12. T. Miya, Y. Teranuma, T. Hosaka, and T. Miyashita, "Ultra Low-Loss Single-Mode Fibre at 1.55 μm," *Electronics Letters*, vol. 15, pp. 106–108, 1979.

13. T. Li, "Structures, Parameters, and Transmission Properties of Optical Fibers," *Proceedings of the IEEE*, vol. 68, pp. 1175–1180, 1980.

14. D. Marcuse, "Curvature Loss Formula for Optical Fibers," *Journal of the Optical Society of America*, vol. 66, pp. 216–220, 1976.

15. K. Petermann, "Microbending Loss in Monomode Fibres," *Electronics Letters*, vol. 12, pp. 107–109, 1976.

16. W. A. Gambling, H. Matsumura, and C. M. Ragdale, "Field Deformation in a Curved Single-Mode Fibre," *Electronics Letters*, vol. 14, pp. 130–132, 1978.

17. K. Petermann, "Fundamental Mode Microbending Loss in Graded Index and W Fibres," *Optical and Quantum Electronics*, vol. 9, pp. 167–175, 1977.

18. A. J. Harris and P. F. Castle, "Bend Loss Measurements on High Numerical Aperture Single-Mode Fibers as a Function of Wavelength and Bend Radius," *IEEE Journal of Lightwave Technology*, vol. LT-4, pp. 34–40, 1986.

19. D. Marcuse, "Microdeformation Losses of Single-Mode Fibers," *Applied Optics*, vol. 23, pp. 1082–1091, 1984.

20. D. Marcuse, "Microbending Losses of Single-Mode, Step-Index and Multimode, Parabolic Index Fibers," *Bell System Technical Journal*, vol. 55, pp. 937–955, 1976.

21. S. Hornung and N. J. Doran, "Monomode Fibre Microbending Loss Measurements and Their Interpretation," *Optical and Quantum Electronics*, vol. 14, p. 359, 1982.

22. D. Marcuse, *Theory of Dielectric Optical Waveguides*. Academic Press, New York, 1974, pp. 126–131.

23. K. Petermann, "Theory of Microbending Loss in Monomode Fibres with Arbitrary Refractive Index Profile," *Archiv fuer Electronik und Uebertragungstechnik*, vol. 30, pp. 337–342, 1976.

5

Dispersion in Step Index Fibers

The information-carrying capacity of any communication system is limited by a number of factors that arise from nonideal behavior in the system components. At the transmitting end, the source will have limitations in attributes such as power and modulation rate. At the receiving end, the detection system will be subject to limitations in speed and detectable signal power—the latter being related to the presence of noise in the detection and amplification processes. Finally, the transmission channel itself will further limit the system capacity as a result of signal degradation arising from losses and dispersion. Losses in the channel subtract from the already limited source power, resulting in possible problems in detecting the weak received signal. Dispersion in the channel distorts any finite bandwidth signal, leading again to detection problems. In optical fiber communication systems, dispersion and loss in the fibers, rather than source and receiver characteristics, are responsible for the ultimate limitations in signal bandwidths that can be transmitted over long distances.

A medium exhibits *chromatic dispersion* if the propagation constant, β, for a wave (or single guided mode) within it varies nonlinearly with frequency. Signal distortion arising from group velocity dispersion will occur as the frequency components of the signal propagate with different velocities; they thus emerge from the medium with a changed time relationship. As will be shown, the severity of this effect is directly proportional to the curvature of an ω–β plot for the medium, measured by the second derivative of β with respect to ω. In an optical waveguide, this effect arises from two mechanisms: (1) from material refractive index variation with wavelength (change in n_1 and n_2 with λ) and (2) from waveguide effects, arising from changes in mode confinement with wavelength. Waveguide dispersion can also be described by changes in the ray angle of the mode with wavelength. The latter effect, as was seen in Chapter 2, will occur in the absence of any material dispersion. The interplay between material and waveguide dispersions in optical fibers has led to a number of successful ways of minimizing net dispersion over a specified wavelength range by using special refractive index profiles.

5.1. GAUSSIAN PULSE PROPAGATION IN A DISPERSIVE MEDIUM

Of particular interest in optical fiber communications is the distortion imposed on a pulse of light as it propagates through a dispersive channel. The simplest case involves the modulation of the output of a monochromatic source of frequency ω_0 by a gaussian pulse envelope. Since the source is monochromatic, the pulse is *transform-limited*, meaning the frequency spectrum is obtained through a Fourier transform of the pulse shape alone. Following the treatment of Marcuse [1], the electric field at $z = 0$ is expressed as

$$E(0, t) = E_0 \exp\left[-\tfrac{1}{2}(t/T)^2\right] \exp(j\omega_0 t) \qquad (5.1)$$

Associated with this pulse is an amplitude spectrum given by its Fourier transform:

$$
\begin{aligned}
E(0, \omega) &= \frac{1}{2\pi} \int_{-\infty}^{\infty} E_0 \exp\left[-\tfrac{1}{2}(t/T)^2\right] \exp[-j(\omega - \omega_0)t]dt \\
&= \frac{E_0 T}{\sqrt{2\pi}} \exp\left[-\tfrac{1}{2}T^2(\omega - \omega_0)^2\right]
\end{aligned}
\qquad (5.2)
$$

Each spectral component of (5.2) is propagated over distance z in the medium, assuming propagation constant $\beta(\omega)$. The pulse amplitude at z is found through the inverse Fourier transform of (5.2) after propagation:

$$E(z, t) = \int_{-\infty}^{\infty} E(0, \omega) \exp[j(\omega t - \beta(\omega)z)]d\omega \qquad (5.3)$$

It is assumed that β exhibits a smooth variation with frequency, allowing its approximation as the first three terms in a Taylor series:

$$
\begin{aligned}
\beta(\omega) &\approx \beta|_{\omega_0} + (\omega - \omega_0) \left.\frac{d\beta}{d\omega}\right|_{\omega_0} + \tfrac{1}{2}(\omega - \omega_0)^2 \left.\frac{d^2\beta}{d\omega^2}\right|_{\omega_0} \\
&= \beta_0 + (\omega - \omega_0)\beta_1 + \tfrac{1}{2}(\omega - \omega_0)^2\beta_2
\end{aligned}
\qquad (5.4)
$$

It is seen that $\beta_1 = 1/v_{g0}$, where $v_{g0} = d\omega/d\beta|_{\omega_0}$ is the group velocity evaluated at the carrier frequency. Substituting (5.4) into (5.3) and evaluating the integral leads to the final expression for the pulse after propagation through distance z:

$$E(z,t) = \underbrace{\frac{E_0[1 - j(\Delta\tau/T)]^{1/2}}{[1 + (\Delta\tau/T)^2]^{1/2}}}_{A} \underbrace{\exp\left[\frac{-(t - \tau_{g0})^2}{2[T^2 + (\Delta\tau)^2]}\right]}_{B}$$

$$\cdot \underbrace{\exp\left[\frac{j(\Delta\tau/T)(t - \tau_{g0})^2}{2[T^2 + (\Delta\tau)^2]}\right]}_{C} \underbrace{\exp[j(\omega_0 t - \beta_0 z)]}_{D} \qquad (5.5)$$

where the *pulse spread* function, $\Delta\tau$, is defined as:

$$\Delta\tau \equiv \frac{\beta_2 z}{T} \qquad (5.6)$$

The character of the transmitted pulse can be understood by examining terms A through C in (5.5), which modulate the original carrier frequency term, D. Term A is the pulse amplitude, which is seen to have diminished from the starting value by an amount that is governed by the dispersive term, β_2. Term B represents the pulse envelope. It is observed that the peak of the pulse reaches the position z after a time $\tau_{g0} = \beta_1 z$, representing the pulse *group delay* evaluated at frequency ω_0. From term B it is also seen that the halfwidth of the pulse power (from the peak to the $1/e$ position) has increased from T to a new value given by

$$T' = [T^2 + (\Delta\tau)^2]^{1/2} \qquad (5.7)$$

It is apparent from (5.7) that an optimum initial pulse width exists for which the output width is minimized. By differentiating (5.7) with respect to T and using (5.6), it is found that this optimum width is given by $T_{opt} = \sqrt{\beta_2 z}$.

Term C in (5.5) represents a frequency modulation that is imposed on the pulse. The instantaneous frequency, ω', is found by taking the time derivative of the exponents in terms C and D:

$$\omega' = \omega_0 + \frac{\Delta\tau(t - \tau_{g0})}{T(T')^2} \qquad (5.8)$$

It is seen that in the case of positive dispersion ($\beta_2 > 0$), the frequency at a fixed position z increases linearly with time (positive linear chirp). If, on the other hand, β_2 is negative, the pulse frequency will decrease with time as it passes an observer at z (negative chirp).

The real part of $E(z,t)$ in (5.5) is plotted in Fig. 5.1b for three different values of the ratio $\Delta\tau/T$. From (5.6), it is seen that increasing this ratio is

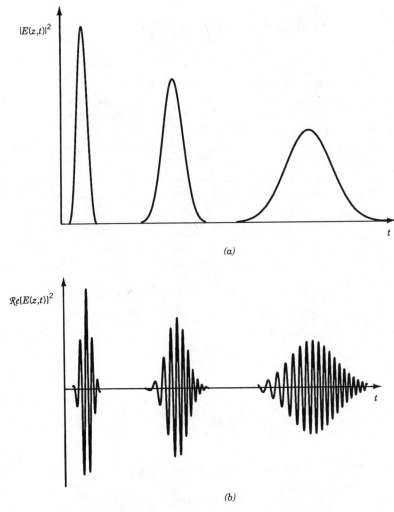

Figure 5.1. Plots of (a) the magnitude squared and (b) the real part of (5.5) for $\Delta\tau/T = 0.5, 2.0$, and 5.0.

equivalent, for example, to increasing the propagation distance of a pulse through a channel with a given amount of chromatic dispersion. The chirping behavior is observed as $\Delta\tau/T$ increases. The pulses are normally observed as intensity envelopes (in which the phase information is absent). The pulse intensities (proportional to $|E(z,t)|^2$) that correspond to the field envelopes of Fig. 5.1b are shown in Fig. 5.1a.

The pulse-broadening and chirping behavior can be further understood by considering the pulse power spectrum as it relates to an ω–β diagram for the medium. The power spectrum, from (5.2), is

$$|E(0,\omega)|^2 = \left| \frac{E_0 T}{\sqrt{2\pi}} \right|^2 \exp[-T^2(\omega - \omega_0)^2] \qquad (5.9)$$

This function has a halfwidth at $1/e$ given by $\Delta\omega = 1/T$ on either side of the center frequency, ω_0. Figure 5.2 shows an $\omega-\beta$ diagram for a medium exhibiting positive dispersion (concave downward) on which these frequencies are plotted. The group velocity at a given frequency is the slope of the $\omega-\beta$ curve at that frequency; the group delay, τ_g, is just the reciprocal of that value multiplied by the propagation distance, z. With this in mind, it is seen that the downward curvature of the plot indicates that positive chirp will be imposed on a pulse whose frequency content lies in the region of the curve that is shown. The pulse spread can be determined by taking the difference between the group delay at frequency ω_0 and that at $\omega_0 \pm \Delta\omega$ (ω_2 or ω_1). Considering ω_0 and ω_2, the group delay difference is

$$\frac{z}{v_g \big|_{\omega_2}} - \frac{z}{v_g \big|_{\omega_0}} = z \left(\frac{d\beta}{d\omega} \bigg|_{\omega_2} - \frac{d\beta}{d\omega} \bigg|_{\omega_0} \right) \qquad (5.10)$$

where

$$\frac{d\beta}{d\omega} \bigg|_{\omega_2} \approx \frac{d\beta}{d\omega} \bigg|_{\omega_0} + \frac{d^2\beta}{d\omega^2} \bigg|_{\omega_0} \left(\frac{1}{T} \right) \qquad (5.11)$$

Terms of second order or higher are assumed negligible. Substituting (5.11) into (5.10) results in

$$\frac{z}{v_g \big|_{\omega_2}} - \frac{z}{v_g \big|_{\omega_0}} = z \frac{d^2\beta}{d\omega^2} \bigg|_{\omega_0} \left(\frac{1}{T} \right) = \frac{\beta_2 z}{T} = \Delta\tau \qquad (5.12)$$

Thus the pulse spread obtained previously by the Fourier transform (5.5) can be determined directly from the $\omega-\beta$ curve using the frequencies corresponding to the halfwidth of the power spectrum. Again, this assumes that the $\omega-\beta$ curve can be accurately represented by the three-term Taylor expansion.

In practice, it is more likely to encounter pulses that have excess bandwidth; that is, the bandwidth exceeds that obtained by a Fourier transform of the pulse shape. A gaussian pulse with excess bandwidth can be modeled by modifying (5.1) to read

$$E(0,t) = E_0(t)\exp\left[-\tfrac{1}{2}(t/T)^2\right]\exp(j\omega_0 t) \tag{5.13}$$

where $E_0(t)$ is the complex amplitude of a polychromatic source, which in general exhibits random fluctuations in amplitude and phase. This behavior is typical of partially coherent light sources such as LEDs and semiconductor lasers.

The simple Fourier transform process presented earlier cannot be used to determine the effects of a dispersive channel on the pulse of (5.13) unless $E_0(t)$ is precisely known. Instead, a statistical treatment is used in which the dispersive propagation effects on the pulse and spectrum autocorrelation functions are analyzed. This method is possible since $E_0(t)$ is a stationary random process, meaning that its autocorrelation, $\langle E_0(t)E_0(t')\rangle$, depends only on $t - t'$. In addition, the power spectral density of the source is assumed gaussian, having full-width $2\Delta\omega_s$:

$$|E_0(\omega - \omega_0)|^2 = \frac{P_0}{\pi^{1/2}\Delta\omega_s}\exp\left(\frac{-(\omega - \omega_0)^2}{(\Delta\omega_s)^2}\right) \tag{5.14}$$

The details of the analysis are presented in [2]. The result is an expression for the ensemble average (pulse-to-pulse) power of the channel output pulses:

$$\overline{P(z,t)} = \frac{P_0 T^2}{[T^4 + (\beta_2 z)^2(1 + T^2(\Delta\omega_s)^2)]^{1/2}}\exp\left(\frac{-T^2(t - \beta_1 z)^2}{[T^4 + (\beta_2 z)^2(1 + T^2(\Delta\omega_s)^2)]}\right) \tag{5.15}$$

In the special case of transform-limited pulses, $\Delta\omega_s = 0$, and (5.15) then exhibits the same temporal and spatial dependence as the square of the magnitude of (5.5). The effect of dispersion on the pulse with excess bandwidth is to broaden it with distance, yielding a pulse at position z having width

$$T' = \left(T^2 + \frac{\beta_2^2 z^2}{T^2}(1 + T^2(\Delta\omega_s)^2)\right)^{1/2} = [T^2 + (\Delta\tau)^2]^{1/2} \tag{5.16}$$

It is convenient to define an effective bandwidth, using (5.16), which combines the source and pulse envelope contributions:

$$\Delta\omega_{\text{eff}} \equiv \left(\frac{1}{T^2} + (\Delta\omega_s)^2\right)^{1/2} \tag{5.17}$$

so that the pulse spread becomes

$$\Delta\tau = \beta_2 z \, \Delta\omega_{\text{eff}} \tag{5.18}$$

In many cases, the source bandwidth is much greater than that associated with the pulse envelope, such that $T^2(\Delta\omega_s)^2 \gg 1$. Thus $\Delta\omega_{\text{eff}} \approx \Delta\omega_s$. If in addition $\beta_2 z \gg T^2$, the initial pulse duration and shape will have minimal influence on the output pulse characteristics.

5.2. MATERIAL DISPERSION

Propagation in a bulk material is considered next, in which dispersion is completely accounted for by the variation of refractive index with frequency. The group delay of a small "spectral packet" centered at frequency ω is given by the propagation distance, z, divided by the group velocity at that frequency:

$$\tau_g = z \, \frac{d\beta}{d\omega} = z \, \frac{d(nk_0)}{d\omega} = z \, \frac{d\lambda}{d\omega} \, \frac{d}{d\lambda} \, (nk_0) \tag{5.19}$$

where n is the refractive index, and λ is the free-space wavelength, given by $\lambda = 2\pi c/\omega$. τ_g can be expressed in terms of wavelength (instead of frequency) through $d\lambda/d\omega = -2\pi c/\omega^2$, from which we obtain

$$\tau_g = \frac{-2\pi c z}{\omega^2} \left(k_0 \frac{dn}{d\lambda} + n \frac{dk_0}{d\lambda} \right) = \frac{z}{c} \left(n - \lambda \frac{dn}{d\lambda} \right) \tag{5.20}$$

From this, the group velocity is expressed as

$$v_g = \frac{c}{n - \lambda(dn/d\lambda)} = \frac{c}{N} \tag{5.21}$$

where the *group index* of the material is defined as

$$N \equiv n - \lambda \frac{dn}{d\lambda} \tag{5.22}$$

The group index provides an alternative to the ω–β curve as a way of thinking about group velocity variation with wavelength. N at a given wavelength will be the speed of light, c, divided by the slope of the ω–β curve at the corresponding frequency. The diagram in Fig. 5.2 shows a decreasing slope with frequency, thus indicating that N will decrease as wavelength increases. A material with this characteristic exhibits *normal group velocity dispersion*.

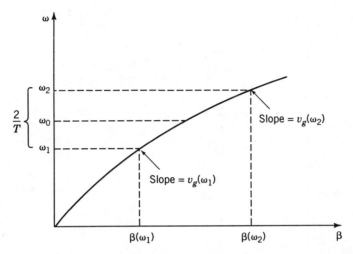

Figure 5.2. ω–β diagram for a material that exhibits normal group velocity dispersion.

Experimentally measured index data can be fitted analytically using the Sellmeier formula for the refractive index, derived in Chapter 1, which is valid at frequencies far from the material resonances:

$$n^2 - 1 = \sum_{j=1}^{p} \frac{A_j \lambda^2}{\lambda^2 - \lambda_j^2} \tag{5.23}$$

The summation is taken over all the resonances. For silica, a three-term Sellmeier equation is typically used, accounting for the resonances in the ultraviolet and infrared:

$$n^2 - 1 = \frac{A_1 \lambda^2}{\lambda^2 - \lambda_1^2} + \frac{A_2 \lambda^2}{\lambda^2 - \lambda_2^2} + \frac{A_3 \lambda^2}{\lambda^2 - \lambda_3^2} \tag{5.24}$$

Equation (5.24) is plotted in Fig. 5.3a for pure silica and for 13.5% germania-doped silica, using the empirically determined Sellmeier coefficients given in Table 5.1 [3,4]. Using (5.24) in (5.22), the group index can be calculated as a function of wavelength. Figure 5.3b shows group index plots that correspond to the materials of Fig. 5.3a. The results show that the group indices for both materials minimize at wavelengths near $\lambda = 1.3 \, \mu m$. These are also the points at which $dN/d\lambda = 0$, which means that group dispersion is zero at these wavelengths. This important feature of silica, along with the fact that losses are low in this wavelength region, is responsible for the choice of near-infrared wavelengths in the vicinity of 1.3 μm for use in fiber-based communication systems.

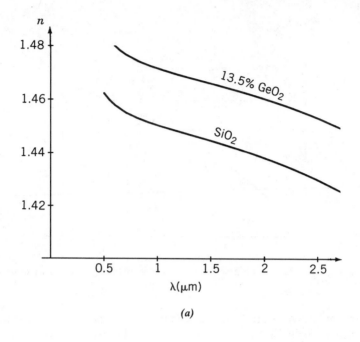

(a)

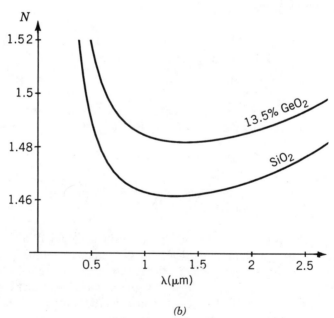

(b)

Figure 5.3. (a) Refractive index and (b) group index curves for two glass compounds, calculated using the data in Table 5.1.

Table 5.1 Sellmeier Parameters Used in Fig. 5.3

Material	A_1	λ_1	A_2	λ_2	A_3	λ_3
SiO$_2$ [3]	0.6961663	0.0684043	0.4079426	0.1162414	0.8974794	9.896161
GeO$_2$:SiO$_2$ [4]	0.711040	0.064270	0.451885	0.129408	0.704048	9.425478

Wavelengths are in μm.

This in turn has resulted in the tremendous interest (and success) in the development of optical sources and detectors that operate at these wavelengths.

The zero group dispersion point for any substance is the wavelength at which an inflection point occurs in the refractive index curve. This can be observed from the relation between Figs. 5.3a and 5.3b. It is also apparent that the inclusion of dopants will shift the zero dispersion wavelength slightly. By combining material dispersion with dispersion arising from the waveguide structure, a value for the zero dispersion wavelength can result that is shifted substantially. The final value, known as λ_0, is an important specification for any single-mode fiber. The parameters that influence the position of λ_0 will be discussed at length in Chapter 6.

From Fig. 5.3b, it is seen that two important regions of the group index curves occur. At wavelengths less than λ_0, group index decreases with increasing wavelength, which means that spectral components of longer wavelengths will travel faster than those of shorter wavelengths. This region is known as that of *normal* group dispersion. For wavelengths greater than λ_0, the opposite behavior of group velocity with wavelength is seen. Consequently, this part of the spectrum is associated with *anomalous* group dispersion.

Earlier it was shown how dispersion relates to the change in group delay with frequency. It is useful to define the *dispersion parameter*, $D(\lambda)$, through

$$D(\lambda) \equiv \frac{dt_g}{d\lambda} = \frac{1}{c}\frac{dN}{d\lambda} \tag{5.25}$$

where $t_g = \tau_g/z$ is the group delay over a unit distance (typically 1 km). From (5.19), $t_g = d\beta/d\omega$, and so

$$D(\lambda) = -\frac{2\pi c}{\lambda^2}\frac{d^2\beta}{d\omega^2} \tag{5.26}$$

At the center wavelength, λ_m, the second derivative in (5.26) evaluates at ω_0, and so the dispersion becomes $D(\lambda_m) = -2\pi c\beta_2/\lambda_m^2$.

Using (5.18), the spread of a pulse whose spectrum is centered on λ_m is determined:

$$\Delta\tau = -\Delta\lambda_{\text{eff}}D(\lambda_m)z \tag{5.27}$$

where $\Delta\lambda_{\text{eff}} = \Delta\omega_{\text{eff}}\lambda_m^2/(2\pi c)$. D is usually expressed in units of picoseconds/ nanometer-kilometer (ps/nm-km), meaning the group delay spread in picoseconds per nanometer of source bandwidth per kilometer of propagation distance. The use of $D(\lambda)$ in characterizing dispersion in fibers is a convenient practice that has become standard.

Group delay and dispersion in a bulk material can be determined analytically by evaluating $dn/d\lambda$ and $dN/d\lambda$ using the Sellmeier formula (5.24). These derivatives are

$$\frac{dn}{d\lambda} = -\frac{\lambda}{n}\left(\frac{A_1\lambda_1^2}{(\lambda^2 - \lambda_1^2)^2} + \frac{A_2\lambda_2^2}{(\lambda^2 - \lambda_2^2)^2} + \frac{A_3\lambda_3^2}{(\lambda^2 - \lambda_3^2)^2}\right) \tag{5.28}$$

$$\frac{dN}{d\lambda} = -\lambda\frac{d^2n}{d\lambda^2}$$

$$= -\frac{N}{n}\frac{dn}{d\lambda} - \frac{4\lambda^3}{n}\left(\frac{A_1\lambda_1^2}{(\lambda^2 - \lambda_1^2)^3} + \frac{A_2\lambda_2^2}{(\lambda^2 - \lambda_2^2)^3} + \frac{A_3\lambda_3^2}{(\lambda^2 - \lambda_3^2)^3}\right) \tag{5.29}$$

where the Sellmeier coefficients are known for the material composition under study. In many cases, data may exist for compositions that bracket the desired one. Excellent results have been obtained by using coefficients derived by linear interpolation using the known data [5]. For example, the Sellmeier coefficients for silica doped with 6.5% germania would have values about midway between those shown in Table 5.1.

5.3. GROUP DELAY IN STEP INDEX FIBERS

This section is concerned with the evaluation of group delay in a fiber and its interpretations. In particular, the first derivative, $d\beta/d\omega$, yields the group delay over a unit distance through the fiber; it is immediately useful in quantifying distortion effects that occur in multimode fibers as a result of group delay differences among the modes. The computation also is the obvious first step in evaluating the second derivative, thus yielding β_2 and ultimately $D(\lambda)$.

To begin, it is necessary to account for dispersive effects that arise from waveguiding as well as material properties. All such properties are included in the fiber propagation constant, β, which is in turn expressed through the b parameter, introduced in Chapter 3:

$$b = \frac{\beta^2 - n_2^2k_0^2}{n_1^2k_0^2 - n_2^2k_0^2} = \frac{w^2}{V^2} \tag{5.30}$$

Solving the above for β results in

$$\beta = n_2 k_0 (1 + 2\Delta b)^{1/2} \approx \underbrace{n_2 k_0}_{A} + \underbrace{n_2 k_0 \Delta b}_{B} \qquad (5.31)$$

In the above, term A contains only material properties, embodied in the refractive index, n_2. Term B exhibits waveguide properties that are expressed in Δ and in b, which are also material dependent. It is thus impossible to fully isolate the material and waveguide dependences in the propagation constant, unless one of these is zero. In taking derivatives of (5.31), such assignments in the final results are typically made for convenience. Specifically, so-called material and waveguide dispersion quantities are identified, but each of these in fact contains contributions from both effects, as will be shown.

To facilitate the evaluation of the first derivative, it is assumed as usual that $\Delta \ll 1$; this enables the normalized index difference to be approximated as $\Delta \approx (n_1 - n_2)/n_2$. Since every term in (5.31) is frequency dependent, its differentiation is a laborious process. It is made easier, however, by appropriate choices of variables. Specifically, differentiations are performed with respect to λ and with respect to V, expressed as $V = (2\pi a/\lambda)n_2\sqrt{2\Delta}$. Use is made of the relations $d/d\omega = -(\lambda^2/2\pi c)d/d\lambda$, as before, and

$$\frac{d}{d\omega} = \frac{\lambda V}{2\pi c}\left(1 - \frac{\lambda}{n_2}\frac{dn_2}{d\lambda} - \frac{\lambda}{2\Delta}\frac{d\Delta}{d\lambda}\right)\frac{d}{dV} \qquad (5.32)$$

Equation (5.32) is used when differentiating b with ω, since b is most conveniently expressed as a function of V. The form of db/dV is further expressed as

$$\frac{db}{dV} = \frac{1}{V}\left(\frac{d(bV)}{dV} - b\right) \qquad (5.33)$$

Using the above forms, the wavelength-dependent group delay over a unit distance is determined:

$$t_g(\lambda, V) \equiv \frac{d\beta}{d\omega} = \frac{1}{c}\{N_1 A(V) + N_2[1 - A(V)] + N_2\Delta[A(V) - b]\} \qquad (5.34)$$

where N_1 and N_2 are the group indices of the core and cladding materials, defined through (5.22), and where

$$A(V) \equiv \frac{1}{2}\left(\frac{d(bV)}{dV} + b\right) \qquad (5.35)$$

For a single-mode fiber, it can be shown (Problem 3.8) that $A(V)$ is the fraction of the LP_{01} mode power that is carried in the core. This leads to an interesting physical interpretation of (5.34) in a single-mode fiber. Since the relative magnitude of the third term in (5.34) (proportional to Δ) is small, the group delay is primarily composed of the weighted sum of group delays in the core and cladding materials. The weighting is determined by the fraction of power that resides in the two regions. It would thus be expected that group delay will increase as more of the power is confined in the core. This is in fact always the case provided the frequency is held constant. The migration of power into the core of a given fiber as frequency increases would also lead to increasing group delay, provided the frequency range lies in the normal dispersion regime ($\lambda < \lambda_0$). For wavelengths larger than λ_0, the group indices decrease as frequency is raised, thus leading to an overall decrease in group delay, even though power moves into the core. A more general way of summarizing the effect is to say that the *effective group index* of the fiber is determined primarily by the weighted sum of the core and cladding group indices, as demonstrated in (5.34).

In multimode fibers, the primary signal distortion mechanism is the feature of differing group delays among the modes, where material dispersion plays a much lesser if not negligible role. This modal effect can be isolated by assuming no material dispersion, such that $N_1 = n_1$ and $N_2 = n_2$. As a result, (5.34) becomes

$$t_g = \frac{n_2}{c}\left(1 + \Delta\frac{d(bV)}{dV}\right) \qquad (5.36)$$

The function $d(bV)/dV$ is plotted in Fig. 5.4 for selected LP modes [6]. Evident in the figure is the fact that for a given value of V, higher order modes generally have larger group delays and thus have slower group velocities. Also evident is that for LP modes for which l is 0 or 1, $d(bV)/dV = 0$ at cutoff, meaning that the group delay per unit distance for these modes at cutoff is n_2/c. This behavior is physically meaningful when considering the power confinement results of Chapter 3 that are summarized in Fig. 3.10. At cutoff, most of the power in LP_{0m} and LP_{1m} modes resides in the cladding, where, in the absence of material dispersion, group velocity would be equal to c/n_2. As frequency increases, a greater percentage of the power in these modes resides in the core region. Figure 5.4 suggests (correctly) that $d(bV)/dV$ approaches unity for all modes as V approaches infinity. At this stage, the power in any mode resides almost entirely within the core and ray trajectories are essentially parallel to the z axis. One would thus expect a group delay of n_1/c, a result that is in fact demonstrated by (5.36). LP modes for which l is greater than 1 exhibit a substantial degree of confinement in the core, even at cutoff. Partly as a result of this, group delay for these modes is n_1/c or greater.

The inclusion of material dispersion enables (5.34) to be recast into a form

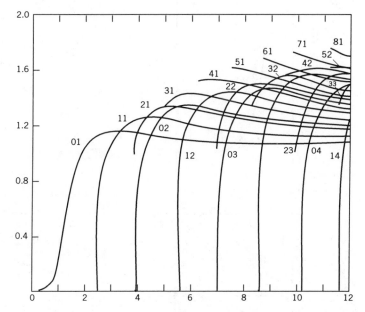

Figure 5.4. Plots of the parameter $d(Vb)/dV$ for various LP modes as functions of V [6].

similar to (5.36):

$$t_g = \frac{N_2 \Delta}{c} (1 + y/4) \frac{d(bV)}{dV} \qquad (5.37)$$

where an additive term involving $d\Delta/d\lambda$ has been neglected. The *profile dispersion* parameter, y, is defined as:

$$y \equiv -\frac{2n_2}{N_2} \frac{\lambda}{\Delta} \frac{d\Delta}{d\lambda} \qquad (5.38)$$

Figure 5.5 shows plots of Δ, $d\Delta/d\lambda$, and y for a fiber constructed from the materials of Fig. 5.3. At transmission wavelengths, the magnitude of y is typically on the order of 0.2 or less. This means that the inclusion of material dispersion leads to some slight modification of the curves of Fig. 5.4 for a given choice of fiber materials. The curves as shown are nevertheless useful in providing good estimates of group delay differences for any multimode fiber.

For multimode fibers, pulse spreading can be estimated from the curves of Fig. 5.4 by computing the group delay difference between the two modes with the largest group delay spacing at a given value of V. Note that this extreme difference becomes larger as V increases and reaches a maximum of $n_2 \Delta/c$ over a unit distance as V becomes very large. This simple result leads to a widely

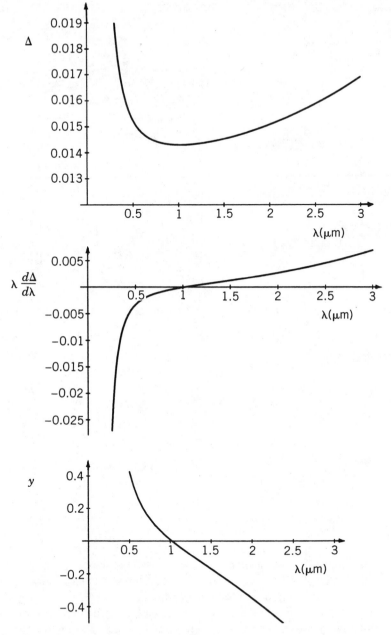

Figure 5.5. Plots of Δ, $\lambda\, d\Delta/d\lambda$, and y for a fiber composed of 13.5% GeO$_2$-doped silica (n_1) and fused silica (n_2). Index functions given by (5.24) were used with the data in Table 5.1.

used approximation formula for the pulse spread in multimode fibers:

$$\Delta\tau \approx \frac{n_2 \Delta}{2c}\, z \tag{5.39}$$

In the multimode case, pulse bandwidth is not at issue, since the calculated group delay differences between the modes would be the same regardless of bandwidth. Only the center frequency (which determines V) is important. Chromatic broadening will occur within the individual modes, but the magnitude of this is almost always much less than the broadening associated with intermodal delay differences, even in the bimoded case.

Numerous attempts have been made to produce simple numerical expressions for $d(bV)/dV$ for single-mode fibers that provide good curve fits. Of these, the most accurate is obtained from the Rudolf–Neumann approximate form of b given in Chapter 3 (Eq. (3.81)):

$$\frac{d(bV)}{dV} \approx 1.3060 - \frac{0.9920}{V^2} = 1.3060 - 0.1715\left(\frac{\lambda}{\lambda_c}\right)^2 \tag{5.40}$$

where λ_c is the cutoff wavelength for the LP_{11} mode. The error in this formula is less than 2% over the range $1.5 < V < 2.4$ [7].

As a final point, it is tempting to try to relate group velocity in a dielectric guide to the core ray velocity projection along the z axis, as is done with success in hollow metallic waveguides. This method works in the latter case because the wave power is totally confined within the metal structure and there is no material dispersion. In dielectric guides, the presence of material dispersion, along with the fact that power resides in the cladding as well as in the core, complicates the analysis and may lead to erroneous results if ray trajectories in the core are considered alone.

5.4.   GROUP DISPERSION IN SINGLE-MODE FIBERS

The group dispersion parameter, $D(\lambda)$, is found through the second derivative of β with ω. Building on the previous results, this is accomplished by differentiating the group delay (5.34) once with respect to λ. The result will lead to the evaluation of D for any mode, distinguished by its particular b function. Of primary interest, however, is the group dispersion of the LP_{01} mode, since, as previously stated, chromatic broadening effects are of little consequence in the multimode case; the D functions for the higher order modes are thus of minor interest.

The above derivative is evaluated using the same variable conversions and term groupings (Eqs. (5.32) and (5.33)) that were used in the last section. The result is*

$$D(\lambda) = D_m(\lambda) + D_w(\lambda) + D_p(\lambda) \tag{5.41}$$

where D_m, D_w, and D_p are referred to as the *composite material, waveguide,* and *profile* dispersions, respectively. Using term groupings that are similar to those in [8], we have

$$D_m(\lambda) = \frac{1}{c} \left(\frac{dN_1}{d\lambda} A(V) + \frac{dN_2}{d\lambda} [1 - A(V)] \right) \tag{5.42}$$

$$D_w(\lambda) = -\frac{N_2^2 \Delta}{n_2 c\lambda} V \frac{d^2(bV)}{dV^2} \tag{5.43}$$

$$D_p(\lambda) = -\frac{N_2^2 \Delta}{n_2 c\lambda} \frac{y}{2} \left(1 + \frac{y}{8}\right) \left(V \frac{d^2(bV)}{dV^2} + \frac{d(bV)}{dV} - b\right) \tag{5.44}$$

where $A(V)$ and y are defined in (5.35) and (5.38), respectively.

The composite material dispersion (5.42) is so-named because it contains the weighted sum of material dispersions in the core and cladding regions [8]. As was true for the group delay (5.34), the weighting is determined by the fractions of mode power in those two regions, given by $A(V)$. Thus waveguiding effects are implicit in D_m, even though the actual dispersion mechanism is in the variation of material group index with wavelength. The waveguide dispersion term (5.43) is referred to as such because of its dependence on the second derivative of bV with respect to V. This term would survive in the absence of material dispersion, even though it contains material properties as embodied in the indices, as well as in b and V. Finally, D_p is termed the profile dispersion, from its dependence on y, or $d\Delta/d\lambda$.

Plots of $D(\lambda)$ for the core and cladding materials of Fig. 5.3 are shown in Fig. 5.6. By (5.42), the composite material dispersion curve for a fiber constructed of these materials will lie somewhere between the two given curves, at a position determined by $A(V)$. Assuming data for the material dispersions are given, the composite dispersion can be calculated using an approximate analytic form for $A(V)$. The latter is found through the previously given approximations for b

---

*This is a lengthy and tedious algebraic exercise that requires much care in keeping track of terms and their proper combination. An additional material dispersion term on the order of Δ and a profile dispersion term involving $[V d^2(bV)/dV^2 - d(bV)/dV + b]$ have been neglected.

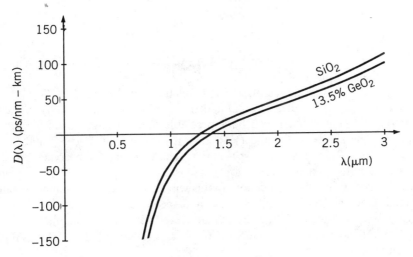

Figure 5.6. Plots of $D(\lambda)$ for the glass compounds of Fig. 5.3.

(Eq. (3.81)) and $d(bV)/dV$ (Eq. (5.40)). Using these in (5.35), the result is

$$A(V) \approx 1.306 - \frac{1.138}{V} = 1.306 - 0.473\left(\frac{\lambda}{\lambda_c}\right) \qquad (5.45)$$

where the range of validity is that of the original formulas: $1.5 < V < 2.4$.

In providing estimates of dispersion, a few approximations are possible, which simplify the waveguide and profile dispersions without significantly compromising the accuracy. First, the index variation with wavelength is ignored when evaluating D_w. This means that $N_2 \approx n_2$ and that $d^2(bV)/dV^2$ is obtainable from existing curves or approximation formulas that apply to any fiber. As a result, (5.43) becomes

$$D_w(\lambda) \approx -\frac{n_2\Delta}{c\lambda} V \frac{d^2(bV)}{dV^2} \qquad (5.46)$$

Second, the profile dispersion can be estimated assuming $y \ll 1$ and, again, $N \approx n$. Equation (5.44) thus becomes

$$D_p \approx \frac{n_2}{c} \frac{d\Delta}{d\lambda}\left(V \frac{d^2(bV)}{dV^2} + \frac{d(bV)}{dV} - b\right) \qquad (5.47)$$

The net pulse spread for a pulse of center wavelength, λ_m, is now calculated through

$$\Delta\tau = -\Delta\lambda_{\text{eff}}[D_m(\lambda_m) + D_w(\lambda_m) + D_p(\lambda_m)]z \qquad (5.48)$$

with D_m, D_w, and D_p as given in (5.42) to (5.44), or in (5.42) along with the approximate forms, (5.46) and (5.47).

Figure 5.7 shows curves of selected functions, including $d^2(bV)/dV^2$, for the LP_{01} mode. As a rule, waveguide dispersion will be substantially less than material dispersion, but it can be appreciable at wavelengths near $1.3\,\mu$m in silica. In the single-mode case ($V < 2.4$), it is seen from Fig. 5.7 and (5.46) that D_w for LP_{01} will always be negative. The fact that D_m becomes positive at wavelengths greater than $1.3\,\mu$m demonstrates the possibility of canceling material and waveguide dispersions in this wavelength range, such that the pulse spread, given by (5.48), would be zero at a specified wavelength (see Problem 5.8). *Dispersion-shifted* fibers are manufactured using this principle.

A simple approximation to the $V\,d^2(bV)/dV^2$ function in Fig. 5.7 was found by Jeunhomme [10]:

$$V\,\frac{d^2(bV)}{dV^2} \approx 0.080 + 0.549(2.834 - V)^2$$

$$= 0.080 + 0.549\left(2.834 - 2.405\,\frac{\lambda_c}{\lambda}\right)^2 \qquad (5.49)$$

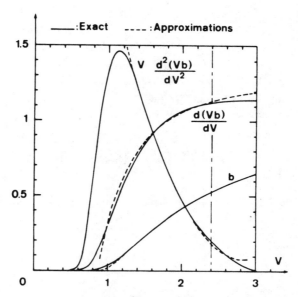

Figure 5.7. Plots of b, $d(Vb)/dV$, and $V\,d^2(Vb)/dV^2$ for the LP_{01} mode. Dotted curves show the approximations to the exact results, given by (3.81), (5.40), and (5.49) [9]. Reprinted from Ref. 9, p. 25 by courtesy of Marcel Dekker Inc.

The error in this approximation is less than 5% over the range $1.3 < V < 2.4$ [10].

A physical interpretation of waveguide dispersion can be seen by examining the idealized $\omega-\beta$ plot shown in Fig. 5.8. In this plot material dispersion is zero, and so the nonlinear change in β with respect to ω results from the change in mode confinement as frequency is varied. At low frequencies (low V) the mode power is mostly in the cladding, and so phase velocity and group velocity (assuming no material dispersion) will both be approximately n_2/c. At very high frequencies (large V) the mode is tightly confined to the core and will thus propagate at phase and group velocities of about n_1/c. This behavior is shown in the $\omega-\beta$ curve, which migrates between the two asymptotes whose slopes are the two velocities, n_1/c and n_2/c.

The group velocity dispersion is as usual found from the second derivative of β with respect to ω. The curve of Fig. 5.8 is seen to exhibit an inflection point (at which $\beta_2 = 0$). This corresponds to the point at which the function $V\, d^2(bV)/dV^2$ reaches zero (at $V = 3$, from Fig. 5.7). One would expect (correctly) that β_2 (and hence D_w) at a specified wavelength would increase with Δ; increasing Δ would have the effect of increasing the angle between the asymptotes in Fig. 5.8 and thus would increase the curvature of the $\omega-\beta$ plot. In addition, the *group delay* arising from waveguide effects will be inversely proportional to the slope of the $\omega-\beta$ curve. Group delay will thus reach a *maxi-*

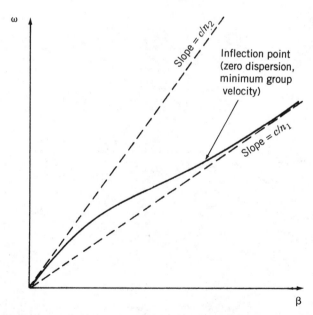

Figure 5.8. Qualitative $\omega-\beta$ diagram for a single-mode fiber, showing the effects of waveguide dispersion only. The zero dispersion point (inflection) is seen to occur where group velocity is minimized.

mum at the inflection point, since the slope minimizes there (recall that for material dispersion, group delay was found to *minimize* at the zero dispersion point). This fact is also seen from Fig. 5.7, where the group delay parameter, $d(bV)/dV$, maximizes at the point where the waveguide dispersion parameter, $V\,d^2(bV)/dV^2$, reaches zero. In summary, waveguide dispersion can be said to result from changes in the power confinement of the mode, in the presence of the two refractive indices.

Plots of D_m, D_w, D_p, and D as calculated from (5.42), (5.46), and (5.47) are shown in Fig. 5.9 for a single-mode fiber having typical parameter values. From the figure it is first observed that the influence of profile dispersion is relatively weak. In addition, the dispersion shifting effect of the waveguide dispersion is apparent. Specifically, a net dispersion curve is produced that is nearly a replica of the material dispersion, but it is shifted to longer wavelengths while being slightly flattened. The methods involved in producing more significant dispersion shifting and flattening will be considered in Chapter 6.

It is common to characterize the combined effects of material, waveguide, and profile dispersion into a net group index parameter for the fiber, N_f, such that

$$D(\lambda) = \frac{1}{c}\,\frac{dN_f}{d\lambda} \tag{5.50}$$

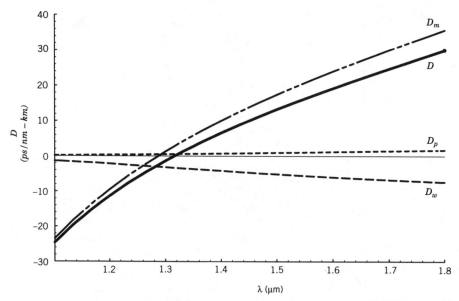

Figure 5.9. Plots of the various dispersion contributions as calculated from (5.42), (5.46), and (5.47), using the approximate formulas (3.81), (5.40), (5.45), and (5.49). Fiber parameters were $\lambda_c = 1.2\ \mu$m, and $\Delta = 0.003$. The refractive index data were obtained by interpolating the Sellmeier data of Table 5.1.

The importance of N_f is that it is readily evaluated experimentally by performing group delay measurements using pulses as described in Section 5.6. Fiber dispersion data are usually given as values of D at specified wavelengths.

Finally, it will be noted that the term "positive" or "negative" group dispersion here refers to the sign of $D(\lambda)$. The roles are reversed when considering dispersion as described by β_2, since $D(\lambda)$ and β_2 are of opposite sign as (5.26) indicates. Consequently, in the general literature, one should be aware of which of the two parameters is being considered when the sign of the dispersion is stated.

5.5. CUBIC DISPERSION

The previous discussion has assumed that chromatic dispersion is quadratic; that is, only the first three terms in the Taylor expansion for $\beta(\omega)$ are used (Eq. (5.5)). The zero dispersion wavelength was defined as the value of λ at which β_2 vanishes. If this occurs, then terms of higher order in (5.5) may be important. To account for this, the cubic term in the Taylor expansion is included, so that (5.5) becomes

$$\beta(\omega) \approx \beta_0 + (\omega - \omega_0)\beta_1 + \tfrac{1}{2}(\omega - \omega_0)^2\beta_2 + \tfrac{1}{6}(\omega - \omega_0)^3\beta_3 \quad (5.51)$$

where $\beta_3 \equiv d^3\beta/d\omega^3|_{\omega_0}$. In a material characterized by index $n(\lambda)$, β_2 and β_3 can be expressed in terms of wavelength through

$$\beta_2 = \frac{\lambda^3}{2\pi c^2}\frac{d^2n}{d\lambda^2} \quad (5.52)$$

$$\beta_3 = \frac{-\lambda^4}{4\pi^2 c^3}\left(3\frac{d^2n}{d\lambda^2} + \lambda\frac{d^3n}{d\lambda^3}\right) \quad (5.53)$$

Plots of (5.52) and (5.53) in fused silica are shown in Fig. 5.10.

As before, a polychromatic source modulated by a gaussian envelope of width T is assumed. The effect of the cubic term is to broaden the pulse while distorting its shape so that it is no longer gaussian. Figure 5.11 shows the effect on a gaussian pulse of initial width T in a channel of length z for the cases in which the dispersion is either purely quadratic or purely cubic [2]. When cubic dispersion dominates, the pulse power envelope is observed to be nonsymmetric and to exhibit oscillations on the trailing edge. The oscillations become more rapid as β_3 increases; they would appear on the pulse leading edge if β_3 were negative. Increasing the quadratic dispersion partially restores the gaussian pulse shape but does so at the expense of increased broadening.

Analysis of the ensemble average pulse power at the output of a channel having both quadratic and cubic dispersion leads to a measure of the pulse

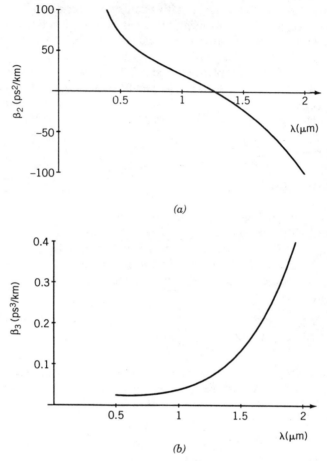

Figure 5.10. Plots of (a) β_2 and (b) β_3 from (5.52) and (5.53), using index data for SiO_2 from Table 5.1.

broadening by way of the variance of the optical power, or square of the rms pulse width: $\sigma^2 = T^2/2 = \overline{t^2} - \overline{t}^2$. This analysis results in a modified expression for the pulse spread appearing in (5.8), which takes the form [2]

$$\Delta\tau = \Delta\omega_{\text{eff}}\left[(\beta_2 z)^2 + \left(\frac{\beta_3 z}{2}\right)^2(\Delta\omega_{\text{eff}})^2\right]^{1/2} \tag{5.54}$$

The oscillatory behavior of the pulse shown in Fig. 5.11c can be understood through the following exercise. The group delay over a unit distance, t_g, can be calculated by differentiating (5.51) with respect to ω. This is done for the two cases of purely quadratic dispersion ($\beta_3 = 0$) and purely cubic dispersion

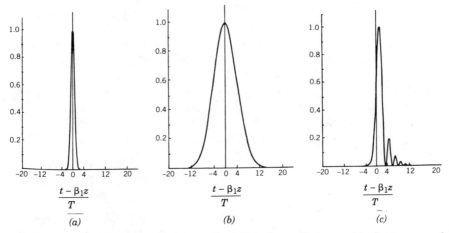

Figure 5.11. Gaussian input pulse (a) to a fiber of length z, and output pulses for the cases of purely (b) quadratic dispersion and (c) purely cubic dispersion [2].

($\beta_2 = 0$) to yield

$$t_{g2} = \beta_1 + (\omega - \omega_0)\beta_2 \tag{5.55a}$$

$$t_{g3} = \beta_1 + \tfrac{1}{2} (\omega - \omega_0)^2 \beta_3 \tag{5.55b}$$

Equation (5.55a) indicates that the effect of quadratic dispersion is to introduce a linear change in group delay with frequency (linear chirp). The effect of cubic dispersion, as shown in (5.55b), is to establish equal group delays for *pairs* of frequencies. The relative phasing between the two frequencies of a given pair will not be the same in all pairs. As a result, the final pulse shape will be influenced by the constructive and destructive interference that will occur between pairs of frequencies, leading to the oscillatory character of Fig. 5.11c.

5.6. SYSTEM CONSIDERATIONS AND DISPERSION MEASUREMENT

Consider a pulse characterized by electric field $E(t)$, which propagates in either a single-mode or multimode fiber. At sufficiently low optical powers, the fiber will exhibit linear behavior with respect to the field, such that the output pulse field can be related to that of the input pulse through the convolution

$$E_{\text{out}}(t) = \int_{-\infty}^{+\infty} h_e(t - \tau)E_{\text{in}}(\tau)d\tau \tag{5.56}$$

where $h_e(t)$ is the complex impulse response of the fiber. Included in h_e are effects that would relate to multimode group velocity differences, mode coupling, and phase shifts associated with chromatic dispersion. h_e will also depend on fiber length. Its measurement would yield all the information needed on the dispersion properties of a given fiber.

Optical measurements almost always involve measurements of power by means of a square-law detector. As a result, it is desirable to be able to express (5.56) in terms of measurable input and output pulse power quantities:

$$P_{out}(t) = \int_{-\infty}^{+\infty} h_p(t - \tau)P_{in}(\tau)d\tau \tag{5.57}$$

where $h_p(t)$ is the impulse response of the fiber associated with power. At first, one would suspect that (5.57) is not valid, since the fiber should logically be assumed nonlinear with power since it is linear with electric field. In fact, (5.57) *is* valid for special cases in which, for example, the source is either monochromatic or of bandwidth that greatly exceeds that associated with the pulse envelope. In these cases (in either single-mode or multimode fiber), the fiber is said to operate in the *quasilinear* regime, and (5.57) can be used with excellent accuracy. The details on the conditions for quasilinearity are presented in [11]. An exercise involving one special case is presented in Problem 5.11.

A useful form of $h_p(t)$ for gaussian pulses can be deduced from the pulse envelope term (*B*) of (5.5):

$$h_p(t) = \exp[-t^2/(\Delta\tau)^2] \tag{5.58}$$

The fiber transfer function is then found from the Fourier transform of (5.58):

$$H_p(\omega) = \int_{-\infty}^{\infty} h_p(t)\exp(-j\omega t)dt = H_p(0)\exp\left(\frac{-\omega^2(\Delta\tau)^2}{4}\right) \tag{5.59}$$

Determining $H_p(\omega)$ enables the evaluation of the *fiber bandwidth*, $\Delta\omega_b$. This is defined as the halfwidth at half-maximum of $H_p(\omega)$.

The effect of finite bandwidth on system performance is to limit the data rate that can be transmitted over a given distance. For data transmission in the form of pulses using binary (on–off) modulation, pulses may spread into adjacent "zero" bit slots, resulting in intersymbol interference. Moreover, they lose amplitude as (5.5) shows, resulting in a higher bit-error rate, since the signal-to-noise ratio at the receiver decreases. Intersymbol interference can be removed to an extent by filtering at the receiver [12], but the drop in pulse amplitude must be compensated by an increase in average power (e.g., at the transmitter) by an amount known as the dispersion *power penalty*.

A method of estimating the power penalty is to use the transfer function (5.59); this is carried out in [13]. The highest frequency of interest in the binary signal will be that of the "dotting" pulse sequence (10101010), which corresponds to radian frequency $\omega = \pi B$, where B is the bit rate. The attenuation of this frequency component is compensated by increasing the average power by the factor $1/H(\pi B)$. By using (5.59), and letting $H(0) = 1$, the power penalty in decibels is

$$P_D = -10 \log_{10} \left[\exp \left(-\frac{(\Delta \tau)^2 (\pi B)^2}{4} \right) \right] \tag{5.60}$$

Using $\log_{10}(x) = 0.434 \ln(x)$, the above reduces to

$$P_D \approx 10.7 (\Delta \tau B)^2 = 21.4 (\sigma_\tau B)^2 \tag{5.61}$$

where σ_τ is the *rms pulse spread*, given by $\sigma_\tau = \Delta \tau / \sqrt{2}$. For a power penalty of 1 dB, (5.61) yields $\sigma_\tau B \approx 1/\sqrt{21}$. Related to this result is the widely used criterion in avoiding excessive dispersion penalties:

$$\sigma_\tau B \leq \tfrac{1}{4} \tag{5.62}$$

Next, consider a fiber length, L, in which the dispersion is mostly quadratic. In this case, the pulse spread is given by (5.25), and so (5.62) becomes $\sigma_\lambda D(\lambda_m) LB \leq \tfrac{1}{4}$, where σ_λ is the rms spectral width, given by $\Delta \lambda_{\text{eff}} / \sqrt{2}$. This result leads to the expression for the maximum allowable bit-rate × distance product for a *dispersion-limited* transmission link:

$$(B \times L)_{\max} = \left[4 \sigma_\lambda D(\lambda_m) \right]^{-1} \tag{5.63}$$

The $B \times L$ product associated with losses can be found using (4.4) (see Problem 4.1). In low-loss cases, the latter result is usually significantly larger than the value determined from (5.63), so that dispersion in fact forms the primary limitation in low-loss links.

The measurement of dispersion in a fiber is an important part of its characterization. Several methods exist, the choice depending on issues such as the type of dispersion to be measured (related to the type of fiber) and the accuracy required. Resulting from the measurements are parameters such as fiber bandwidth for a specified length, or the dispersion, D. The measurement techniques can be grouped roughly into two categories—frequency domain and time domain. A few examples are described below.

One method is to obtain the impulse response of a fiber by deconvolving measured input and output pulse shapes. The experiment uses a known input pulse, which is transform-limited for best results. After propagation through the fiber, the broadened pulse envelope is measured and its Fourier transform, $P_{out}(\omega)$, is calculated. The Fourier transform of the impulse response (the fiber transfer function) is then found as the ratio of the output and input pulse transforms:

$$H_p(\omega) = FT\{h_p(t)\} = \frac{P_{out}(\omega)}{P_{in}(\omega)} \qquad (5.64)$$

where FT denotes Fourier transform. Inverse-Fourier transforming $H_p(\omega)$ yields $h_p(t)$, having width $\Delta\tau$. This method is usually (and most easily) applied to multimode fibers, in which chromatic dispersion is essentially negligible. It is thus not necessary to be concerned with the precise spectral characteristics of the source pulse, as would be necessary with single-mode fibers. Furthermore, relatively long pulses, whose shapes and widths are readily measured, can be used in the multimode case.

Pulse broadening and fiber bandwidth are necessarily length dependent, as was already shown. In the simplest case, the dependence is linear, such that the product of bandwidth and length of a given fiber is a constant. On the other hand, significant departures from linearity can be seen in multimode fibers, owing to effects of differential mode loss and mode coupling. In the latter case, diffusion of power from one mode to another can occur, leading to a distribution of power that is nonuniform among the modes, or which at least may differ from that which existed at the input. As a result, bandwidth may vary with length in a manner that is not linear. Added complications arise when several fibers are end-to-end coupled to form a concatenated length. These include mode power redistribution as a result of differing index profiles among the component lengths, as well as mode conversion that occurs at the joints.

The concatenation problem is addressed by expressing the bandwidth dependence on length through the formula

$$\Delta f(L) = \left(\frac{L}{L_0}\right)^\gamma \Delta f_0 \qquad (5.65)$$

where L is the total length of the assembled fiber, and L_0 is the length of each section having bandwidth Δf_0. γ is the so-called bandwidth–length extrapolation factor, assuming values within the range $0.5 \le \gamma \le 1.0$. It has been found possible to predict γ for arbitrary initial mode power distributions [14].

The range of γ can be understood intuitively by considering the mechanism of mode diffusion. Transfer of power between modes will occur most efficiently from a given mode to the adjacent lower and higher order modes. There is thus a tendency for power to migrate from one mode to the next, which has the effect of reducing the overall time spent in the lowest and highest order modes. The group delay differences are thus effectively averaged to an extent, leading to less pulse broadening. The fiber bandwidth thus decreases with length in a manner that is less than linear.

Measurement of single-mode fibers is not generally done by the deconvolution method, because of the previously mentioned source characterization problem and because the pulse widths involved are usually too short to allow convenient shape and width evaluation. Instead, time domain techniques that involve group delay measurements can be used. Specifically, pulses having different carrier frequencies are individually propagated through a length of test fiber. The *differences* in the pulse group delays are measured for the pulses as frequency is varied, resulting in a measurement of $dN_f/d\lambda$. Dispersion, D, is then readily found using (5.50).

Frequency domain techniques form another class of dispersion measurement methods. These are usually based on measurements of phase shifts that occur between a modulated wave that propagates through a length of fiber and a reference. A slow sinusoidal modulation of radian frequency Ω is applied to a continuous wave (cw) laser source, which is input to a length of fiber. The modulation envelope will propagate at the group velocity evaluated at the carrier frequency and will thus experience a wavelength-dependent group delay, $t_g(\lambda)$, over a unit length. The phase of the detected signal at the fiber output is compared to that of the laser modulator as the carrier frequency is varied. The latter can be accomplished by substituting different lasers.

The phase difference between the envelopes at two different carrier frequencies at distance z can be expressed as $\Delta\phi = \Omega\,\Delta t$, where Δt is the difference in the respective group delays. In turn, the group delay change can be related to the change in wavelength through $\Delta t = (dt_g/d\lambda)z\Delta\lambda$, from which $\Delta\phi = \Omega(dt_g/d\lambda)z\Delta\lambda$. The result is a measurement of dispersion, $D(\lambda)$, through

$$D(\lambda) = \frac{dt_g}{d\lambda} = \frac{\Delta\phi}{\Omega\,\Delta\lambda z} \qquad (5.66)$$

An alternate approach to performing measurements in frequency domain is to observe the detected signal level from the fiber output as the source modulation frequency is varied over a wide range, thus mapping out the $H_p(\omega)$ function. The fiber bandwidth is measured directly by noting the frequency at which the detector voltage falls to one-half its value at zero modulation. The physical basis for the roll-off in H_p as frequency increases can be explained as follows. In multimode fibers, increasing the source modulation frequency increases the extent to which the power carried in different modes will destructively interfere, thus reducing the net output power. This is because group delay differences over the test length will be on the order of a modulation cycle—thus allowing the possibility of destructive interference between modes as modulation frequency increases. In single-mode fibers, the effect of increasing the modulation frequency is to increase the level of destructive interference between the spectral components of the modulated carrier. The latter will propagate with different group delays and can thus destructively interfere provided group delay differences are on the order of the modulation period.

PROBLEMS

5.1. Consider the propagation of a transform-limited gaussian pulse in a dispersive medium in which the propagation constant can be written as

$$\beta \approx \beta_0 + (\omega - \omega_0)\beta_1 + \tfrac{1}{2}(\omega - \omega_0)^2\beta_2$$

Chirp is introduced on the pulse, where the instantaneous frequency, ω', is given by (5.11).

a) Determine the rate of frequency change with time, $d\omega'/dt$, at fixed position z. Assume that broadening is sufficient to allow the approximation $T' \approx \Delta\tau$.

b) What requirements (pulse characteristics, medium properties, etc.) must be met if the frequency variation with time is to be linear?

c) The chirping rate becomes less as z or β_2 increases. Provide a physical explanation.

5.2. A single-mode fiber is constructed of glass in which the core index variation with wavelength obeys approximately the Cauchy formula:

$$n(\lambda) = 1.58 + \frac{1.46 \times 10^6}{\lambda^2}$$

where λ is expressed in angstroms (Å).

a) Calculate the phase velocity and the group index for light at 5145 Å.

b) Calculate the group velocity of a light pulse centered at 5145 Å.

c) A 5145 Å light pulse of initial duration 1 ps (full-width at half-maximum, FWHM) and of 20 Å bandwidth is to be transmitted through 1 km of this fiber. Assuming negligible waveguide dispersion and assuming that the intensity is confined mostly to the core, determine the approximate pulse width after transmission. For the calculation, also assume "quasi-plane wave" operation; that is, $\beta \approx n_1 k_0$.

5.3. A transform-limited gaussian pulse of initial width $T = 1$ ps is input to a single-mode fiber of length 10 m. The output pulse is measured to be of width $T' = 2$ ps.

a) A different gaussian pulse, also of 1 ps width, but which has a bandwidth of twice that of the original pulse, is input to the same fiber. Estimate the output pulse width for this case.

    b) A transform-limited gaussian pulse of 2 ps width is input to the fiber. Estimate the output pulsewidth.

    c) The original 2 ps output pulse of the problem statement is repropagated through the same fiber. Estimate the output pulse width and compare with your result of part b.

5.4. The *dispersion length* parameter, L_D, is defined as the distance over which a transform-limited gaussian pulse will broaden to $\sqrt{2}$ times its initial width, T. Referring to (5.7), show that $L_D = T^2/|\beta_2|$. Also determine the distance over which a pulse having source bandwidth $\Delta\omega_s$ will broaden to $\sqrt{2}$ times its initial width. Express that result in terms of L_D.

5.5. A certain material has its dispersion properties characterized by a refractive index function:

$$n(\lambda) = n_0 - n_1\lambda - n_2\lambda^2(\lambda - 3\lambda_0)$$

where n_0, n_1, and n_2 are constants.

    a) Determine and make a sketch of the group index, N, of the material as a function of wavelength.

    b) Consider a transform-limited gaussian pulse of center wavelength λ_m ($\lambda_m < \lambda_0$) and halfwidth at $1/e$ of T. Show that the pulse spread over a distance L in the material will be $\Delta\tau = 3n_2\lambda_m^3 L(\lambda_0 - \lambda_m)/(\pi c^2 T)$, where c is the velocity of light in free space.

    c) Determine the optimum initial pulse width, T_{opt}, such that the output pulse width is minimized.

    d) A second pulse, identical to the first except for center wavelength λ_s, is input to the material simultaneously with the first pulse. Write an equation relating λ_s to λ_m such that the peaks of the two pulses will arrive simultaneously at the output.

5.6. Show that, by a power series expansion of (5.24), the refractive index function of a material can be approximated as

$$n(\lambda) \approx C_1 + C_2\lambda^2 + \frac{C_3}{\lambda^2}$$

where C_1, C_2, and C_3 are constants. Show also that the group index function can be approximated, using the above index, as

$$N(\lambda) \approx C_1 - C_2\lambda^2 + \frac{3C_3}{\lambda^2}$$

5.7. For a material characterized by the expansions of Problem 5.5, determine the zero dispersion wavelength in terms of the given constants.

5.8. For a transform-limited pulse, the optimum input pulse width that minimizes the pulse width at the output of a channel of length z exhibiting quadratic dispersion was shown to be $T_{opt} = \sqrt{\beta_2 z}$.

   a) Determine T_{opt} for a pulse having excess bandwidth that propagates in a medium in which the dispersion is quadratic with no cubic.

   b) Determine T_{opt} for a transform-limited pulse that propagates in a medium in which the dispersion is cubic with no quadratic.

5.9. A train of pulses from a mode-locked semiconductor laser is to be propagated through a Ge-doped silica single-mode fiber for which $\Delta = 2 \times 10^{-3}$ and $V = 1.8$. The laser operates at center wavelength $\lambda = 1.55 \ \mu m$ and the output is of bandwidth $\Delta\lambda = 10$ Å. The pulses are of 10 ps width and their centers are spaced by 1 ns. An acceptable signal at the fiber output for this system is one in which the individual pulse widths (T') are no greater than one-fourth the pulse separation.

   a) Considering the effects of material and waveguide dispersion, calculate the maximum length of fiber through which this signal can be transmitted to yield an acceptable output.

   b) Next, suppose that a different laser is used, having the same output characteristics except for a shorter wavelength. The normalized frequency of the fiber changes to $V = 3.5$, allowing propagation of the LP_{11} mode set in addition to LP_{01}. Discuss the effects that this situation has on the form of the received signal, and provide an estimate of the maximum allowed length.

5.10. A single-mode step index fiber is to be constructed for operation at the minimum loss wavelength of 1.55 μm. The cladding material is to be pure silica and the required value for V is 2.2. Specify approximate values for the radius and refractive index of the core such that the net dispersion will be zero at 1.55 μm.

5.11. Verify that (5.53) is valid for the case of a multimode fiber in which the modes are not coupled. Source bandwidth is very small, such that chromatic dispersion can be neglected. Pulse broadening is thus due to intermodal delay differences only. The procedure is based on the requirement of orthogonality of the modes.

5.12. Transmission through single-mode fiber over long distances usually requires the end-to-end joining of several segments of fiber whose dispersion properties may differ. A *dispersion map* of the link consists of a tabulation of the D and L values for each segment. If cubic dispersion is negligible, it is possible to achieve zero dispersion over the aggregate link by assuring that the sum of the products of D and L for the joined segments is zero. Explain physically how this scheme works. This principle of *dispersion equalization* is discussed at length in ref. 15. Descriptions of passive techniques used to compensate chromatic dispersion are found in refs. 16 and 17.

REFERENCES

1. D. Marcuse, *Light Transmission Optics.* Van Nostrand Reinhold, New York, 1982.

2. D. Marcuse, "Pulse Distortion in Single-Mode Fibers," *Applied Optics*, vol. 19, no. 10, pp. 1653–1660, May 15, 1980.

3. I. H. Malitson, "Interspecimen Comparison of the Refractive Index of Fused Silica," *Journal of the Optical Society of America*, vol. 55, pp. 1205–1209, 1965.

4. J. W. Fleming, "Material Dispersion in Lightguide Glasses," *Electronics Letters*, vol. 14, pp. 326–328, 1978.

5. H.-C. Huang and Z.-H Wang, "Analytical Approach to Prediction of Dispersion Properties of Step-Index Single Mode Optical Fibers," *Electronics Letters*, vol. 17, pp. 202–204, 1981.

6. D. Gloge, "Weakly Guiding Fibers," *Applied Optics*, vol. 10, pp. 2252–2258, 1971.

7. R. A. Sammut, "Analysis of Approximations for the Mode Dispersion in Monomode Fibres," *Electronics Letters*, vol. 15, pp. 590–591, 1979.

8. W. A. Gambling, H. Matsumara, and C. M. Ragdale, "Mode Dispersion, Material Dispersion and Profile Dispersion in Graded-Index Single-Mode Fibres," *Microwaves, Optics and Acoustics*, vol. 3, pp. 239–246, 1979.

9. L. B. Jeunhomme, *Single-Mode Fiber Optics,* 2nd ed. Marcel Dekker, New York, 1990.

10. L. Jeunhomme, "Single-Mode Fiber Design for Long Haul Transmission," *IEEE Journal of Quantum Electronics*, vol. QE-18, pp. 727–732, 1982.

11. S. D. Personick, "Baseband Linearity and Equalization in Fiber Optic Digital Communication Systems," *Bell System Technical Journal*, vol. 52, pp. 1175–1194, 1973.

12. S. D. Personick, "Receiver Design for Digital Fiber Optic Communication Systems" (Parts I and II), *Bell System Technical Journal*, vol. 52, pp. 843–874 and 875–886, 1973.

13. P. S. Henry, R. A. Linke, and A. H. Gnauck, "Introduction to Lightwave Systems," in *Optical Fiber Telecommunications II*, S. E. Miller and I. P. Kaminow, eds. Academic Press, Orlando, FL, 1988.

14. D. A. Nolan, R. M. Hawk, and D. B. Keck, "Multimode Concatenation Modal Group Analysis," *IEEE Journal of Lightwave Technology*, vol. LT-5, pp. 1727–1732, 1987.

15. D. Marcuse, "Equalization of Dispersion in Single-Mode Fibers," *Applied Optics*, vol. 20, pp. 696–700, 1981.

16. L. J. Cimini, Jr., L. J. Greenstein, and A. A. M. Saleh, "Optical Equalization to Combat the Effects of Laser Chirp and Chromatic Dispersion," *IEEE Journal of Lightwave Technology*, vol. 8, pp. 649–659, 1990.

17. Special Mini-Issue on Dispersion Compensation, *IEEE Journal of Lightwave Technology*, vol. 12, pp. 1706–1765, 1994.

6

Special Purpose Index Profiles

The preceding chapters have dealt solely with the step index fiber design. This form, while structurally the simplest, exhibits some performance features that are in many situations less than adequate. For example, in the single-mode case it was found that waveguide dispersion could be used to reduce the overall chromatic dispersion at specified wavelengths that are greater than 1.3 μm by appropriate choice of core index and radius. While this feature is attractive, the usual result is that a large index difference and small core radius are required; these can result in higher scattering losses and can yield mode field radii that are unacceptably small, possibly resulting in higher coupling losses. In multimode fibers, it was found that significant bandwidth limitations occur that arise from group delay differences between modes. The magnitude of this effect can only be reduced by decreasing the index difference, which reduces the power confinement. Finally, the circular symmetry of the fiber cross section precludes any possibility of controlling the polarization of the guided light.

The above shortcomings of the single-step index profile initiated the search for other profiles that would yield better performance. In the multimode case, it was found that, by using a core index that gradually decreases with radius according to a specific function, group velocities for all the modes could nearly be equalized, thus providing decreased signal distortion over that found in the multimode step index case. In single-mode fibers, more complicated profile designs that involve several steps, along with index grading, were found to give superior performance in dispersion control, while operating with large mode field radii and low losses. Other profiles that are not circularly symmetric were found useful in either maintaining the linear polarization of the input light or in allowing only a single polarization to propagate. The description and analyses of fibers having more general index profiles are the topics of this chapter.

6.1. MULTIMODE GRADED INDEX FIBERS

Index profiling in multimode fibers is typically done by producing a gradual monotonic decrease in core index with radius, such that the value at the cladding boundary is equal to the index of the cladding. It will be seen that intermodal group delay differences can be reduced substantially by using a simple core profile function. Since multimode fibers have relatively large core sizes, the *slowly varying index* criterion usually applies:

$$\left| \lambda \, \frac{dn/dx}{n} \right| \ll 1 \tag{6.1}$$

One way to interpret this is to say that the variation in index that occurs over a distance comparable to a wavelength is much less than the average index in that region. With (6.1) satisfied, geometrical optics will be valid—that is, ray paths will be smooth, having no abrupt changes in trajectory at which diffraction could occur. Core sizes of fibers that meet this requirement are relatively large, achieving diameters of about 25 μm and larger.

The most common index function is the power law or *alpha profile* [1]:

$$n(r) = n_1 \left[1 - 2\Delta \left(\frac{r}{a} \right)^{\alpha} \right]^{1/2} \qquad r \le a \tag{6.2}$$

where n_1 is the index on the fiber axis and n_2 is the cladding index. The normalized index difference is defined using these constants as was done in the step index case; that is, $\Delta = (n_1^2 - n_2^2)/2n_1^2$. The profile parameter, α, is in the vicinity of 2 for best reduction of group delay differences among modes. Added design flexibility can be achieved by the so-called multiple-α profile functions [2]:

$$n(r) = n_1 \left[1 - \sum_i 2\Delta_i \left(\frac{r}{a} \right)^{\alpha_i} \right]^{1/2} \qquad r \le a \tag{6.3}$$

where $\Delta = \sum_i \Delta_i$.

The effect on ray trajectories produced by a profile exemplified by (6.2) or (6.3) is shown qualitatively in Fig. 6.1. In the figure, ray 1 represents a mode of lower order than that represented by ray 2. Both rays are meridional in this case, although skew rays are also possible. Two important concepts are associated with this figure. First, a single ray will have a z-directed propagation constant, β, that will be single-valued over the entire ray path. This can be seen by considering the graded index region as composed of a number of concentric cylindrical layers, each of which is of constant refractive index; the index decreases from layer to layer, as radius increases. A ray in such a structure will refract at successive interfaces to form a path that will approach a smooth curve as the layer thicknesses are made smaller. At each refraction, however, the boundary conditions dictate that β must be the same on either side of the interface. Hence β cannot change over the entire ray path. The second concept concerns the possibility of equalizing the optical paths of the two rays shown. Ray 2 is seen to propagate over a longer distance than ray 1. The refractive

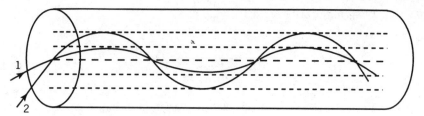

Figure 6.1. Ray paths associated with two modes in a graded index fiber.

index encountered by ray 2, however, is on average lower than that encountered by ray 1. Therefore ray 2 will propagate at a higher average velocity along its path. With appropriate choice of the index function, the "times of flight" for both rays (group delays) can be made equal.

Figure 6.2 shows the components of a ray path in a graded index fiber, in analogy to Fig. 3.3 (Chapter 3). As in the step index case, a three-dimensional boundary value problem in cylindrical coordinates again arises, in which the z variation is assumed to be $\exp(-j\beta z)$. A difference in the approach to be taken here is that the weakly guiding condition is invoked at the outset of the analysis, enabling LP modes to be postulated as solutions of the wave equation. The latter is thus separable into equations for each of the cartesian field components; the presumed single component of $\mathbf{E}$ in the transverse plane is then found by solving the wave equation for that component.

Following the reasoning of Chapter 3, the ϕ component of the propagation constant, β_ϕ, will be equal to l/r, where l is the LP azimuthal mode number, an integer. As a result, the radial propagation constant can be expressed as

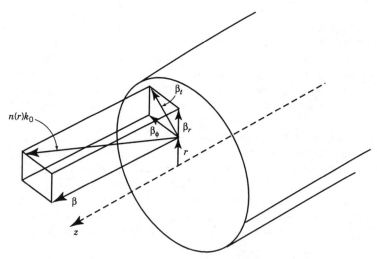

Figure 6.2. Ray geometry in a fiber having radially dependent refractive index.

$\beta_r = [n^2(r)k_0^2 - \beta^2 - l^2/r^2]$. Radial mode number, m, will label the possible values of β_r, which, at a given point, can only assume discrete values. The resulting graded index fiber modes will have ϕ and r variations that are associated with mode numbers l and m; that is, the LP_{lm} modes that arise for the graded index fiber correspond directly to those of the step index fiber. In the present case, the LP_{lm} modes will appear qualitatively similar to their step index counterparts, in that the same numbers of variations in the r and ϕ directions will be seen; the dimensional scaling of the graded index versions, however, will be different. Grading the index will also produce differences in mode confinement and cutoff conditions.

6.2. FIELD ANALYSIS OF MULTIMODE GRADED INDEX FIBERS

The derivation of the mode fields must begin with wave equations that are appropriate for the medium. Since refractive index now varies with position, the familiar forms of the equations for **E** and **H** given by (1.19) and (1.20) are no longer valid, since these were derived under the assumption of constant permittivity. The wave equations are rederived using the methods outlined in Chapter 1, while allowing for the spatial variation in ϵ. The steps of this procedure are left as an exercise (Problem 6.1). The results are

$$\nabla^2 \mathbf{E} + \omega^2 \mu \epsilon \mathbf{E} + \nabla \left(\mathbf{E} \cdot \frac{\nabla \epsilon}{\epsilon} \right) = 0 \qquad (6.4)$$

$$\nabla^2 \mathbf{H} + \omega^2 \mu \epsilon \mathbf{H} + \frac{\nabla \epsilon}{\epsilon} \times \nabla \times \mathbf{H} = 0 \qquad (6.5)$$

Aside from the added complexity, the main feature that distinguishes these wave equations from those of the homogeneous case is that they are not separable into the individual field components. For example, it is possible using (1.19) to solve for one field component, such as E_z, and then find the others using Maxwell's equations, as was done in Chapters 2 and 3. In contrast, on substituting the field vectors into (6.4) and (6.5), the last terms in each (involving $\nabla \epsilon / \epsilon$) would break up into three components, but each of these would be produced from a field component that lies in a different direction. Consequently, three component equations would result for each of (6.4) and (6.5), but each of the six would contain more than one component of **E**. Thus the solution of a single equation for a single component is generally not possible.

Equations (6.4) and (6.5) are separable if the last terms in each are negligible, which is true if the term $\nabla \epsilon / \epsilon$ is very small. This approximation, known as the *small index gradient condition*, is related to (6.1). It is not as well defined, however, since its applicability would depend on the expected spatial rate of change in $\mathbf{E}$ and $\mathbf{H}$, as the last terms in (6.4) and (6.5) imply. Nevertheless, satisfaction of (6.1) is sufficient to invoke the condition of small $\nabla\epsilon/\epsilon$ in all cases, and the third terms in the wave equations can be dropped. Field solutions are assumed of the form

$$\mathbf{E} = \mathbf{E}_0(r,\phi)\exp(-j\beta z), \quad \text{and} \quad \mathbf{H} = \mathbf{H}_0(r,\phi)\exp(-j\beta z) \qquad (6.6)$$

On substitution of the above into the simplified forms of (6.4) and (6.5), and assuming index variation only with radius, the wave equations become

$$\nabla_t^2 \mathbf{E}_0 + [n^2(r)k_0^2 - \beta^2]\mathbf{E}_0 = 0 \qquad (6.7)$$

$$\nabla_t^2 \mathbf{H}_0 + [n^2(r)k_0^2 - \beta^2]\mathbf{H}_0 = 0 \qquad (6.8)$$

where ∇_t^2 is the Laplacian taken with respect to the transverse coordinates, r and ϕ.

With the assumption of LP mode fields as the solutions of (6.7) and (6.8), the single electric field component, E_x, for example, can be found directly from (6.7) (with $\mathbf{E}_0$ replaced by E_x). As in the step index case, the field amplitudes of each mode are expressed in product solution form. Thus E_x becomes the sum of the individual mode fields of the form

$$E_x = \sum_i R_i(r)\Phi_i(\phi)\exp(-j\beta_i z) \qquad (6.9)$$

Substituting one term of (6.9) into (6.7), the wave equation separates to become

$$\frac{d^2 R(r)}{dr^2} + \frac{1}{r}\frac{dR(r)}{dr} + \left(n^2(r)k_0^2 - \beta^2 - \frac{l^2}{r^2}\right)R(r) = 0 \qquad (6.10)$$

and

$$\frac{d^2\Phi(\phi)}{d\phi^2} + l^2\Phi(\phi) = 0 \qquad (6.11)$$

The solution of (6.11) is $\Phi = \cos(l\phi + \delta)$ or $\sin(l\phi + \delta)$.

Although (6.10) resembles Bessel's equation, it is not unless $n^2(r)$ is a constant. Use of the alpha profile (6.2) for $n(r)$ in (6.10) will in fact not lead to closed form solutions except for the case in which $\alpha = 2$, where Laguerre–gaussian or Hermite–gaussian modes result, as discussed in ref. 3 (the special case of the $l = 0$ gaussian mode is considered in Problem 6.4). In the general case (for general values of α), an approach involving approximations is used to solve (6.10), which is accurate provided ray trajectories are smooth, as assumed here. This is the WKB method, invented by Lord Rayleigh, but later applied to quantum mechanics problems by G. Wentzel, H. Kramers, and L. Brillouin.*In the method, the radial function for a given mode is expressed in the form of a single phase term:

$$R(r) = \exp[-jk_0 s(r)] \qquad (6.12)$$

where $s(r)$ is a scalar function that defines a family of constant-phase surfaces. At any position, the ray direction is found by taking the gradient of $s(r)$.

Equation (6.12) is now substituted into (6.10) to obtain

$$k_0^2 (s')^2 + jk_0 \left(s'' + \frac{1}{r} s' \right) - \left(n^2(r)k_0^2 - \beta^2 - \frac{l^2}{r^2} \right) = 0 \qquad (6.13)$$

where s' and s'' are first and second derivatives of s with respect to r. To facilitate the solution of (6.13), $s(r)$ is expanded in powers of $1/k_0$ (powers of wavelength), resulting in

$$s(r) = s_0(r) + \frac{1}{k_0} s_1(r) + \frac{1}{k_0^2} s_2(r) + \cdots \qquad (6.14)$$

At this stage, the *WKB approximation* is invoked, in which only the first two terms of (6.14) are retained. This condition implies that the wavelength is small compared to all physical dimensions of interest. Additionally, the condition is another manifestation of the small index gradient approximation (the variation of n is small over wavelength-order distances). Note also that as $\lambda \to 0$, only the first term in (6.14), $s_0(r)$, survives. This is the *geometrical optics limit*, in which wave propagation is accurately characterized by the use of rays. Substituting the first two terms of (6.14) into (6.13) results in

$$k_0^2 \left[\left(n^2 k_0^2 - \beta^2 - \frac{l^2}{r^2} \right) \frac{1}{k_0^2} - (s_0')^2 \right]$$
$$- k_0 \left[2s_0' s_1' + j \left(s_0'' + \frac{1}{r} s_0' \right) \right]$$
$$- \left[(s_1')^2 + j \left(s_1'' + \frac{1}{r} s_1' \right) \right] = 0 \qquad (6.15)$$

*An earlier user, H. Jeffreys, is sometimes credited by referring to the technique as the WKBJ method.

The above must hold regardless of the value of k_0. Consequently, each of the expressions within square brackets in (6.15) must individually be zero. The equation thus separates into three smaller equations, which can be solved for s_0 and s_1. The first and second bracketed terms in (6.15) produce the following equations:

$$\left(n^2(r)k_0^2 - \beta^2 - \frac{l^2}{r^2} \right) - (s_0')^2 k_0^2 = 0 \qquad (6.16)$$

$$2s_0's_1' + j\left(s_0'' + \frac{1}{r}s_0' \right) = 0 \qquad (6.17)$$

Equation (6.16) can be solved by direct integration to obtain the zeroth order solution:

$$s_0(r) = \pm \frac{1}{k_0} \int_0^r \left(n^2(r')k_0^2 - \beta^2 - \frac{l^2}{r'^2} \right)^{1/2} dr' \qquad (6.18)$$

In zeroth order, the radial function is now $R(r) = \exp[-jk_0s_0(r)]$. This will be oscillatory (implying a traveling wave) as long as $s_0(r)$ is real, which, from (6.18), will be true when

$$[n^2(r)k_0^2 - \beta^2 - l^2/r^2] > 0 \qquad (6.19)$$

This condition will be satisfied over a particular range of r values within the core region. To see this, a graphical method is presented [4,5] in which a plot of $n^2(r)k_0^2 - l^2/r^2$ as a function of radius is shown in Fig. 6.3. A horizontal line, indicating the value of β^2, is added to the plot. As observed in the figure, the range $r_1 < r < r_2$ is the region in which (6.19) is valid. Outside this region, the function $[n^2(r)k_0^2 - \beta^2 - l^2/r^2]$ is less than zero, which means that $s_0(r)$ becomes imaginary. This produces solutions resembling exponentials that decay away from either side of the $r_1 < r < r_2$ oscillatory region. The radii r_1 and r_2 are the *turning points*, or *caustics*. At these locations, the ray reaches its minimum or maximum position in radius. The path of a ray would thus resemble that shown in Fig. 6.4a. In this case, the ray follows a skew path, which remains within the region defined by $r_1 < r < r_2$. When $l = 0$, it is found that only one intersection point occurs between the function $n^2(r)k_0^2$ and β^2 (see Problem 6.2). The result is that the mode solution will be oscillatory over the range $r < r_1$. The mode, having designation LP_{0m}, is thus identified as meridional.

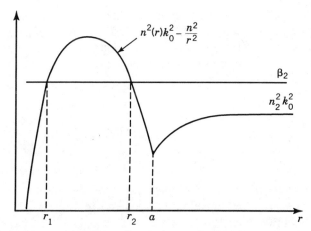

Figure 6.3. Qualitative plots of $n^2(r)k_0^2 - l^2/r^2$ and β^2 (straight line) as functions of radius, showing the region between turning points, r_1 and r_2, where the mode field is oscillatory as dictated by (6.18) and (6.19).

The analysis continues by substituting (6.18) into (6.17) and solving for $s_1(r)$, to obtain

$$s_1(r) = -\frac{j}{4} \ln \left[\frac{r^2}{k_0^2} \left(n^2(r)k_0^2 - \beta^2 - \frac{l^2}{r^2} \right) \right] \qquad (6.20)$$

which is seen to have poles at the two turning points. The WKB approximation is thus not valid in the vicinity of the turning points, so other approximation methods must be used there. Briefly, the procedure is to assume that the function $n^2(r)k_0^2 - l^2/r^2$ is approximately linear between two boundaries that are a small distance on either side of each turning point (Fig. 6.3). Solutions are obtained

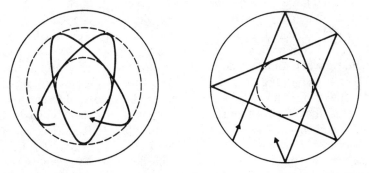

Figure 6.4. Skew ray trajectory as projected in fiber transverse cross sections for the (a) graded index and (b) step index cases [6]. © 1987 Artech House, Norwood, MA.

from the wave equations (Airy functions) that apply to these regions. All solutions must then be connected at the four boundaries (by evaluating amplitude and phase constants) such that one continuous solution across the entire core is obtained. The connection process can be carried out if the following approximate condition is met:

$$\int_{r_1}^{r_2} \left(n^2(r)k_0^2 - \beta^2 - \frac{l^2}{r^2} \right)^{1/2} dr \approx m\pi \tag{6.21}$$

where m, the radial mode number, plays the same role that it had in the slab guide and step index fiber. Equation (6.21) is known as the *WKB quantization rule*. This description is appropriate since, as is apparent in the equation, β can only assume discrete values. The equation is recognized as the graded index equivalent of the transverse resonance condition, since the left-hand side evaluates the accumulated radial phase shift between turning points. This must be performed by integration, rather than by a simple product of β_r with a radial distance, since index varies with radius. The condition becomes exact when l^2 and m are replaced by $l^2 + \frac{1}{4}$ and $m + \frac{1}{4}$, meaning that lower order modes are not precisely described using the WKB method, whereas the accuracy improves for higher order modes. The simplified form expressed by (6.21) is nevertheless adequate in situations involving numerous modes, in that a reasonably accurate and intuitive view of general mode behavior results; this is useful in evaluating and optimizing fiber bandwidth, as will be shown in the next section.

It is seen from (6.21) that, for a given frequency, a larger value of β will correspond to a smaller value of m. This is because the integrand of (6.21) decreases, and the turning points move closer together, as seen in Fig. 6.5. From that figure, note that if β is less than n_2k_0, a second region appears in which $R(r)$ is oscillatory. This area extends from a third turning point, r_3, to infinity. Radiation in this region escapes, and the mode is thus leaky. The confined portion of the mode, between r_1 and r_2, escapes to the outer region by means of evanescent coupling between the two regions. In this phenomenon, referred to as "optical tunneling," some of the power penetrates the "barrier" existing between r_2 and r_3. As β decreases further, the outer barrier eventually vanishes, and even partial confinement is not possible.

Changing the azimuthal mode number, l, produces a new set of allowed β values. This is seen in Fig. 6.6, in which curves for two values of l are shown; a single value of β is shown for each l. The β values are seen to be close, and, in some cases, they could be equal. In the figure, β_a has mode numbers l_a and m_a, while l_b and m_b are associated with β_b. Since $l_a > l_b$, it follows from (6.21) that $m_b > m_a$. As l increases, the part of the function $n^2(r)k_0^2 - l^2/r^2$ that lies inside the core region becomes narrower and decreases in overall amplitude. This leads to a decrease in the number of solutions to (6.21)—or fewer values of m.

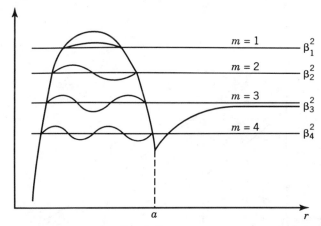

Figure 6.5. Qualitative plot of $n^2(r)k_0^2 - l^2/r^2$ along with four values of β^2, corresponding to four values of radial mode number, m. The $m = 4$ mode is partially confined and is identified as a leaky wave.

The way in which the modes are grouped is shown in Fig. 6.7. Here, individual modes are indicated as dots, having coordinates m and l. Curves of constant β are shown, which pass near some mode positions and intersect others. At a given position, β increases toward the origin, in a direction normal to the constant-β curve at that point. The maximum β value will occur at the $l = 0$, $m = 1$ mode position. At a given frequency, the minimum value of β will occur along the contour that is furthest from the origin.

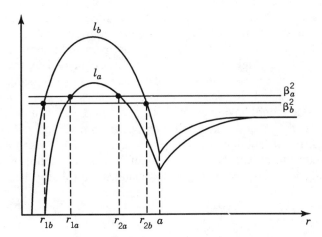

Figure 6.6. Plots as per Fig. 6.5, showing curves for two values of l and two values of m, such that the resulting values of β are nearly equal.

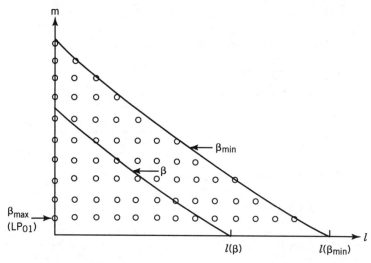

Figure 6.7. Representation of modes of designation LP_{lm} that occur at a given frequency. Those closest to cutoff lie near the curve indicating the minimum β value. The maximum value of β will be that of the LP_{01} mode, as shown.

6.3. MULTIMODE INDEX PROFILE OPTIMIZATION

Using the results of the last section, it is possible to determine the profile shape that will yield the minimum group delay difference between modes. The procedure is to obtain an approximate expression for β, and hence group delay, as functions of mode number and refractive index profile. The optimum index profile is then found by minimizing the resulting expression for group delay differences between the lowest and highest order modes. This analysis is valid only for cases in which a very large number of modes is present, in which case (6.21) is accurate for most of the modes under consideration.

The first step is to count the modes that are present between a given value of β and β_{max}. The highest order mode in this count will be the one closest to cutoff and will be associated with the value of β that is being considered. The number of modes in each computation becomes the *mode number* of the mode (or mode group) having propagation constant β. Referring to Fig. 6.7, one way to count the modes is to count the dots in each vertical column between β_{max} and β, for all columns between $l = 0$ and $l = l(\beta)$. These totals are then added together to obtain the net number of modes. The number of modes in the column l will be given by the WKB quantization rule for the chosen value of β:

$$m(\beta) = \frac{1}{\pi} \int_{r_1(l)}^{r_2(l)} \left(n^2(r)k_0^2 - \beta^2 - \frac{l^2}{r^2} \right)^{1/2} dr \qquad (6.22)$$

The total number of modes between β and β_{max} is then given by

$$\nu(\beta) = 4 \sum_{l=0}^{l(\beta)} m(\beta) \tag{6.23}$$

The factor of 4 is introduced because there are usually four modes associated with each l and m pair, representing the four possible polarization/orientational configurations of each LP mode (except for LP_{0m}). The configurations for LP_{11}, for example, were shown in Fig. 3.7.

If $l(\beta)$ is assumed very large, the summation can be treated as an integral. Substituting (6.22) into (6.23) results in

$$\nu(\beta) = \frac{4}{\pi} \int_0^{l(\beta)} \int_{r_1(l)}^{r_2(l)} \left(n^2(r)k_0^2 - \beta^2 - \frac{l^2}{r^2} \right)^{1/2} dr \, dl \tag{6.24}$$

The integral is easier to evaluate if the order of integration is reversed. To accomplish this, it is necessary to know how the turning point locations change as a function of l. Turning points occur whenever the integrand of (6.24) is zero. This will occur whenever

$$l = r[n^2(r)k_0^2 - \beta^2]^{1/2} \tag{6.25}$$

This function is plotted qualitatively in Fig. 6.8. From the figure, it is seen that the turning points with maximum separation will be $r = 0$ and $r = r_2(0)$, corresponding to $l = 0$ (meridional ray). Using these and (6.25), the order of integration of (6.24) can be reversed, such that the equation now reads

$$\nu(\beta) = \frac{4}{\pi} \int_0^{r_2(0)} \int_0^{r[n^2(r)k_0^2 - \beta^2]^{1/2}} \left(n^2(r)k_0^2 - \beta^2 - \frac{l^2}{r^2} \right)^{1/2} dl \, dr \tag{6.26}$$

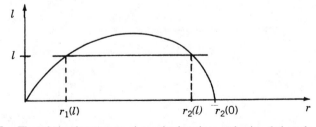

Figure 6.8. The relation between turning point locations and azimuthal mode number, l.

Integrating over l results in

$$\nu(\beta) = \int_0^{r_2(0)} [n^2(r)k_0^2 - \beta^2] r \, dr \tag{6.27}$$

The value of $r_2(0)$ is determined by (6.25) with $l = 0$; that is,

$$n^2(r_2(0))k_0^2 - \beta^2 = 0 \tag{6.28}$$

Consider now the alpha profile, given by (6.2). Substituting this into (6.28) and solving for $r_2(0)$, we obtain

$$r_2(0) = a \left(\frac{n_1^2 - \beta^2/k_0^2}{n_1^2 - n_2^2} \right)^{1/\alpha} = a(1 - b)^{1/\alpha} \tag{6.29}$$

where b is the normalized propagation constant defined in Chapter 3. Substituting (6.29) and (6.2) into (6.27) and carrying out the integration, we find

$$\nu(\beta) = a^2 k_0^2 n_1^2 \Delta \left(\frac{\alpha}{\alpha + 2} \right) (1 - b)^{(\alpha+2)/\alpha} \tag{6.30}$$

The total number of modes is found by determining ν for the minimum possible value of β, which is $n_2 k_0$. When $\beta = n_2 k_0$, $b = 0$, and (6.30) becomes

$$\nu_T = \nu(n_2 k_0) = a^2 k_0^2 n_1^2 \Delta \left(\frac{\alpha}{\alpha + 2} \right) = \frac{V^2}{2} \left(\frac{\alpha}{\alpha + 2} \right) \tag{6.31}$$

It therefore follows that

$$(1 - b) = \left(\frac{\nu}{\nu_T} \right)^{\alpha/(\alpha+2)} = \frac{n_1^2 - \beta^2/k_0^2}{n_1^2 - n_2^2} = \frac{n_1^2 k_0^2 - \beta^2}{2\Delta n_1^2 k_0^2} \tag{6.32}$$

Solving for β results in an expression analogous to (3.79):

$$\beta(\nu) = n_1 k_0 \left[1 - 2\Delta \left(\frac{\nu}{\nu_T} \right)^{\alpha/(\alpha+2)} \right]^{1/2} \tag{6.33}$$

where ν is now considered a fixed number and ν_T is given by (6.31).

Group delay over a unit distance can now be determined as a function of "mode group" number, ν, by using $t_g = 1/v_g = d\beta/d\omega = (1/c)d\beta/dk_0$. Expanding (6.33) into a binomial series and taking the above derivative, we find [7]

$$
\begin{aligned}
t_g = \frac{N_1}{c} & \left[1 + \Delta \left(\frac{\alpha - 2 - y}{\alpha + 2} \right) \left(\frac{\nu}{\nu_T} \right)^{\alpha/(\alpha+2)} \right. \\
& \left. + \frac{\Delta^2}{2} \left(\frac{3\alpha - 2 - 2y}{\alpha + 2} \right) \left(\frac{\nu}{\nu_T} \right)^{2\alpha/(\alpha+2)} + \cdots \right]
\end{aligned}
\tag{6.34}
$$

where N_1 is the group index of the core material at $r = 0$, and y is the profile dispersion parameter, given by

$$
y = -\frac{2n_1}{N_1} \frac{\lambda}{\Delta} \frac{d\Delta}{d\lambda}
\tag{6.35}
$$

Consider now the *difference* in group delays between the lowest and highest order modes ($\nu = 0$ and $\nu = \nu_T$). Using (6.34) and neglecting terms of order higher than Δ^2, the group delay difference will be given by

$$
\begin{aligned}
\Delta t_g & = t_g(\nu = \nu_T) - t_g(\nu = 0) \\
& = \frac{N_1}{c} \left[\Delta \left(\frac{\alpha - 2 - y}{\alpha + 2} \right) + \frac{\Delta^2}{2} \left(\frac{3\alpha - 2 - 2y}{\alpha + 2} \right) \right]
\end{aligned}
\tag{6.36}
$$

Since Δ is usually very small, it is permissible to neglect the term in Δ^2. The first term in (6.36) is reduced to zero when α achieves the optimum value:

$$
\alpha_{\text{opt}} = 2 + y
\tag{6.37}
$$

This important result describes the relationship between the optimum profile shape and the material dispersion, represented by y. With no material dispersion, (6.37) indicates that the simple parabolic profile ($\alpha = 2$) will optimize fiber performance. Using (6.37) in (6.36) and again neglecting the Δ^2 term, it is found that

$$
\Delta t_g = \frac{N_1 \Delta}{c} \left(\frac{\alpha - \alpha_{\text{opt}}}{\alpha + 2} \right)
\tag{6.38}
$$

This states that with $\alpha > \alpha_{\text{opt}}$, the higher order modes will arrive at the output later than the fundamental mode; the reverse is true when $\alpha < \alpha_{\text{opt}}$.

It is thus apparent that a graded index fiber must be designed for a particular wavelength. Deviation from this wavelength will result in increased differences between group delays for the modes, with the resulting increased signal distortion. This effect, embodied in the profile dispersion parameter, y, is minimized if material choices are made such that y varies as little as possible with wavelength. Reasonably good results have been obtained using ternary glasses such as $P_2O_5:GeO_2:SiO_2$.

With $\alpha = \alpha_{opt}$, the second term in (6.36), involving Δ^2, is no longer negligible. Substituting (6.37) into (6.36) results in

$$\Delta t_g(\alpha_{opt}) \approx \frac{N_1 \Delta^2}{2c} \tag{6.39}$$

The values of Δt_g as expressed in (6.38) or (6.39) are used to determine the pulse spread as used in (5.17) through

$$\Delta \tau = \frac{\Delta t_g}{2} z \tag{6.40}$$

A performance comparison can now be made to the step index case by evaluating (6.36) for the case in which $\alpha \to \infty$. Neglecting the Δ^2 term, the result is

$$\Delta t_g(\alpha \to \infty) \approx \frac{N_1 \Delta}{c} \quad \text{(step index)} \tag{6.41}$$

By taking the ratio of (6.39) to (6.41), it is evident that the graded index multimode fiber provides a reduction in pulse spreading over a step index fiber having the same Δ by a factor of $\Delta/2$. Further improvements in performance can be realized by more careful study of (6.36) (Problem 6.7), but there is no known profile that will completely equalize all mode group delays.

6.4. SINGLE-MODE GRADED INDEX FIBERS

The use of continuously varying index profiles in single-mode fibers enables the control of key performance parameters, such as the mode field radius, and losses that arise from scattering and bending. The development of this practice was motivated by the early efforts to modify the dispersive properties of fibers—specifically, to vary the location of λ_0 and to flatten the dispersion as a function of wavelength. The latter have been accomplished by using modified step index profiles, as will be considered in Section 6.5. These designs, however, can lead to increased scattering losses, in addition to mode field radii that are too small (leading to increased coupling loss). It was found that the losses could be substantially reduced by smoothing the index variation in the basic step index designs for dispersion modification. The loss characteristics depend on the index profile in a very complicated way, and many trade-offs exist. Thus some compromise must usually be met in determining the optimum profile shape for a given application.

Index grading may also occur, not by design but as a result of the inherent properties or deficiencies in the manufacturing processes. In the MCVD process, for example, these include the practical difficulty of achieving a perfect step at the core–cladding boundary due to dopant diffusion, and the common index depression on the fiber axis; the latter arises from volatization, or "burnout," of the dopant materials near the core center during the preform collapsing stage. In the VAD process, graded core fibers are in fact the easiest to manufacture, owing to the nature of the process. In particular, the gaussian profile has been found to give good performance, while being readily fabricated using VAD [8].

It was previously shown that the mode field distribution in a single-mode step index fiber will vary in the core as $J_0(ur/a)$ and in the cladding as $K_0(wr/a)$; it was also shown that a gaussian distribution can be used to model this combination of functions by appropriate choice of the mode field radius, r_0. Consider now the effect of slightly departing from the pure step by allowing the transition from core to cladding indices to occur over a small change in radius; this could yield, for example, an index profile that is trapezoidal in shape. Assuming the difference between the two indices is small (weakly guiding case), the mode field will undergo little change in its shape over that of the original step index fiber, aside from a slight increase in its width (mode field radius), accompanied by a decrease in peak amplitude. The field form can in fact be closely approximated using appropriate J_0 and K_0 functions (or a gaussian) that describe the field of *some other* fiber having a step index profile. This *equivalent step index fiber* may have the same cladding index as the graded fiber, but its core index and radius will be different. Once these are determined, the explicit mode field expression for the graded fiber will be known. Other parameters for the graded fiber such as the mode propagation constant, mode field radius, and cutoff wavelength can be found. In addition, splicing and bending losses for a graded fiber can be determined with fairly good accuracy by applying step index models for these losses to the equivalent step fiber. Various methods exist for determining the equivalent step index parameters for a given graded index fiber, either through analysis [9,10], or by measurement [11]. These are readily applicable to all cases in which the graded core index monotonically decreases from the fiber axis, although good results can be obtained for cases in which moderate index depressions occur. A description of one of the analytic methods, that of Snyder and Sammut [9], and some of its key results are presented here.

Consider a graded single-mode fiber having core radius a_g and core index function $n_g(r)$. Under the weakly guiding approximation, the small index gradient condition will hold, even though the core radius is small. Equation (6.7) will thus apply and will assume the following form:

$$\nabla_t^2 E_g(r,\phi) + [n_g^2(r)k_0^2 - \beta_g^2]E_g(r,\phi) = 0 \tag{6.42}$$

where $E_g(r, \phi)$, the transverse field amplitude, is assumed linearly polarized and real. The solution of (6.42) will yield a total field of the form $\mathcal{E}_g = \frac{1}{2}E_g(r, \phi)\exp(-j\beta_g z) + \text{c.c.}$ A *variational* approach can be used to solve the equation. Multiplying (6.42) by E_g and integrating the result over any cross-sectional plane yield

$$\int_{A_\infty} E_g \nabla_t^2 E_g + [n_g^2(r)k_0^2 - \beta_g^2]E_g^2\, da = 0 \qquad (6.43)$$

Since the mode field is independent of ϕ, and since β_g is independent of r and ϕ, (6.43) can be written in the following form [9]:

$$\beta_g^2 = \frac{\displaystyle\int_0^\infty [E_g\nabla_t^2 E_g + n_g^2(r)k_0^2 E_g^2]r\, dr}{\displaystyle\int_0^\infty E_g^2 r\, dr} \qquad (6.44)$$

Equation (6.44) has the property of being *stationary* with respect to changes in *variational parameters* that influence E_g. In other words, small deviations in E_g from its exact form for a given index profile will result in very little change in β_g. This will be true as long as the first partial derivatives of β_g^2 with respect to each of the parameters that can be used to describe E_g (such as mode field radius, u, and w) are zero. The second derivatives may not be zero, however, leading to the conclusion that the correct value of β_g^2 will occur at a minimum, a maximum, or a saddle point on a surface obtained by plotting β_g^2 as a function of the parameters. To simplify the analysis, the number of parameters used is the minimum required to fully describe the field. In the present case, it can be shown through analysis of (6.44) that β_g^2 will be maximized when the correct E_g is found. Various techniques can be used to find the maximum β_g. Among these is the substitution of presumed forms of E_g (trial functions) into (6.44), and then noting the improvement on the resulting β_g value over those obtained from previous substitutions. The *Rayleigh–Ritz* method involves writing the trial function for E_g in the form of a series of functions of the variational parameters. The expansion is substituted into (6.44), resulting in a set of equations that can be solved for the correct parameter values under the condition that all first partial derivatives of the equations with respect to the parameters must be zero. Detailed discussions of the variational method and properties of stationary functions can be found in refs. 12 and 13.

The equivalent step index method involves using a step index fiber field, E_s, as a trial solution of (6.44). In this case, the chosen parameters are the step fiber core radius, a_s, and its normalized frequency, V_s. It will be recalled that these two quantities are sufficient to fully characterize the shape of E_s; this is evident, for example, in the gaussian approximation studies that lead to Eq. (3.87). The analysis procedure can be simplified by first noting that an expression analogous to (6.44) can be written for the step index fiber:

$$\beta_s^2 = \frac{\int_0^\infty [E_s \nabla_t^2 E_s + n_s^2(r)k_0^2 E_s^2]r\,dr}{\int_0^\infty E_s^2 r\,dr} \tag{6.45}$$

where $n_s(r) = n_{1s}$ for $r < a_s$ and is equal to n_2 for $r > a_s$. For (6.45) to represent the equivalent step fiber, it is required that $E_s \approx E_g$. Under this condition, (6.45) can be subtracted from (6.44) to yield a new stationary expression for β_g:

$$\beta_g^2 - \beta_s^2 \approx \frac{\int_0^\infty [n_g^2(r) - n_s^2(r)]k_0^2 E_s^2 r\,dr}{\int_0^\infty E_s^2 r\,dr} \tag{6.46}$$

Again, the correct β_g is found by choosing the values of a_s and V_s such that β_g^2 is maximized in (6.46). Normalized expressions for E_s having the following form are substituted into (6.46):

$$E_s = \begin{cases} J_0(u_s r/a_s)/J_0(u_s) & r \le a_s \\ K_0(w_s r/a_s)/K_0(w_s) & r \ge a_s \end{cases} \tag{6.47}$$

where u_s and w_s can be found for a given V_s by solving the eigenvalue equation for LP_{01}, Eq. (3.48).

Further simplification is needed, since (6.46) explicitly depends on the normalized index differences, Δ_g and Δ_s, which appear in the refractive indices and propagation constants. They are in fact already represented in the V_g and V_s parameters, so their additional appearance is redundant and complicates the analysis.*The core index functions for the step and graded index fibers are expressed in terms of the Δ's as

$$n_s^2(r) = n_{1s}^2[1 - 2\Delta_s f_s(r)] \tag{6.48}$$

---

*The core radii, a_g and a_s, are also contained in V_g and V_s, but their separate appearance as variational parameters is necessary to fully specify the shape of the mode field. Specifying core radius thus means that V is varied only through changes in Δ or in λ.

$$n_g^2(r) = n_{1g}^2[1 - 2\Delta_g f_g(r)] \tag{6.49}$$

where $f_g(r)$ is the given graded index profile function, and where $f_s(r)$ is one for $r > a_s$ and is zero for $r < a_s$. In addition, (3.14) and (3.46) are used to find the Δ dependence in the propagation constants:

$$\beta_{g,s}^2 = \frac{1}{a_{g,s}^2}\left(\frac{V_{g,s}^2}{2\Delta_{g,s}} - u_{g,s}^2\right) \tag{6.50}$$

To remove the Δ_s and Δ_g dependence in (6.46), (6.48) through (6.50) are substituted into (6.46) to yield the following stationary expression for u_g^2 [9]:

$$u_g^2 \approx \left(\frac{a_g}{a_s}\right)^2 (u_s^2 - V_s^2 F_s) + V_g^2 F_g \tag{6.51}$$

where

$$F_{s,g} \equiv \frac{\displaystyle\int_0^\infty f_{s,g}(r)E_s^2 r\,dr}{\displaystyle\int_0^\infty E_s^2 r\,dr} \tag{6.52}$$

Since β_g must be maximized in working with (6.46), (6.51) must be solved for the *minimum* u_g (as (6.50) indicates) by finding appropriate values of a_s and V_s. Note that the other unknown quantities in (6.51), u_s and E_s, depend directly on these parameters.

Results of a general nature can be obtained by applying the equivalent step method to the alpha profile, where $f_g(r) = (r/a_g)^\alpha$ for $r \le a_g$ and is unity for $r \ge a_g$. Solving (6.51) for the minimum u_g as a function of V_g for various α values leads to the curves shown in Fig. 6.9. These results are nearly identical to those resulting from exact numerical solutions of the wave equation; the error increases with V_g, reaching only about 0.5% for V_g on the order of 5 [9].

Using the above method, the equivalent step fiber core index and radius are found to vary as V_g changes, although these changes are very slight over the range of single-mode operation. This behavior is shown in Fig. 6.10, in which equivalent step profiles are shown for an $\alpha = 2$ fiber, for a few values of V_g. In another method [10], in which a gaussian field is assumed, the equivalent step structural parameters do not vary as V changes. In either method, the calculated values of β_g and β_s are found to differ very slightly from one another; this difference approaches zero for graded fibers whose profiles differ slightly from a step profile, as (6.46) indicates.

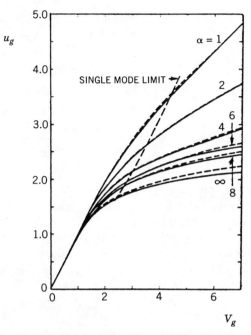

Figure 6.9. Plots of u_g as determined from (6.52) as functions of V_g [9].

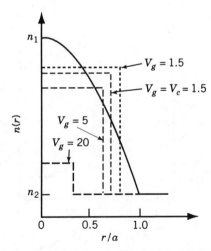

Figure 6.10. Equivalent step profiles for an $\alpha = 2$ fiber [9].

The present method can be used to find the value of V_g at cutoff for the LP_{11} mode for a given alpha profile. The procedure is to assume that the LP_{11} mode field in the graded fiber will be nearly the same as that on some step index fiber. At cutoff, $V_s = u_s = 2.405$ and $V_g = u_g = V_c$. Equation (6.51) thus becomes

$$V_c^2 \approx (2.405)^2 \left(\frac{a_g}{a_s} \right)^2 \frac{1 - F_s}{1 - F_g} \qquad (6.53)$$

The normalized field expressions used in the above are derived from (3.68) and (3.69), from which only the radial variations are used (note that any fiber field exhibits the same ϕ dependence inside and outside the core; thus the inclusion of integration over ϕ in (6.44) would have no effect on the result). The normalized field in the core is then $E_s = J_1(2.405r/a_s)/J_1(2.405)$ for $r < a_s$. Since $w \to 0$ near cutoff, (3.69) is used with the small argument approximation of K_1 (Eq. (A.16) in Appendix A), resulting in $E_s \approx a_s/r$ for $r > a_s$. The step index core radius is varied to determine the minimum value of V_c in (6.53). V_c thus obtained is plotted as a function of α in Fig. 6.11. These results are nearly identical to those of a different procedure involving use of a series expansion

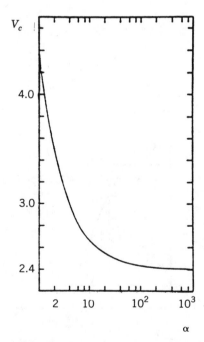

Figure 6.11. V at cutoff for the LP_{11} mode as a function of profile paramater, α [9].

for the field solution [14]. Of particular interest are the values $V_c = 4.38$ for $\alpha = 1$ and $V_c = 3.52$ for $\alpha = 2$. Note that, as expected, V_c approaches 2.405 in the step index limit, that is, as $\alpha \rightarrow \infty$.

The fact that V_c increases as α decreases is indicative of the decreased ability of the graded profile to contain the LP_{11} mode when compared to the step index case for a given core radius, index difference, and wavelength. An approximate formula for V_c was derived, based on this behavior [15]. In it, a *guidance factor*, G, is defined as

$$G = \int_0^{a_g} \frac{n^2(r) - n_2^2}{n_1^2 - n_2^2} \frac{r}{a_g} \, dr \qquad (6.54)$$

Note that G ranges in value from zero to a maximum of $\frac{1}{2}$; the latter value corresponds to the step index case. Using (6.54), the value of V at cutoff is then found through the empirical formula

$$V_c \approx \frac{2.405}{\sqrt{2G}} \qquad (6.55)$$

Using the above, values of V_c for the $\alpha = 1$ and $\alpha = 2$ cases are determined to be 4.17 and 3.40, respectively, which are in error by 4.9% and 3.3%. The error decreases with α and approaches zero as the step index case is approached. The importance of the LP_{11} cutoff condition lies not only in calculating the range of single-mode operation but also in determining the contributions of various loss mechanisms to the fundamental mode. This latter point becomes important, in particular, when formulating designs for dispersion-shifted fiber, to be considered in Section 6.5.

The variational method also provides a way of determining the gaussian mode field parameters that are used to approximate the LP_{01} field. In this application, the trial function of (6.52) is taken to be a gaussian of the form

$$E_p = E_0 \exp[-(r/r_0)^2] \exp(-j\beta z) \qquad (6.56)$$

Whereas the previous trial function, (6.47), is the exact solution of a step index profile, (6.56) is the exact solution for the lowest order mode of an infinite parabolic index profile, having refractive index $n(r) = n_1[1 - 2\Delta(r/a_p)^2]^{1/2}$ ($0 < r < \infty$), where $n(a_p) = n_2$. This can be verified by substituting (6.56) into the wave equation (6.7) in which the above index distribution is used. Resulting are the following expressions for the mode field radius and normalized radial propagation constant for the parabolic profile (Problem 6.4):

$$r_0 = \sqrt{\frac{2}{V_p}}\, a_p \tag{6.57}$$

$$u_p = \sqrt{2V_p} \tag{6.58}$$

where $V_p = a_p k_0 n_1 \sqrt{2\Delta}$.

It is now assumed that an *equivalent parabolic* profile will exist whose mode field, given by (6.56), will closely approximate that of a general fiber index profile having parameters a_g and V_g [9]. Equation (6.51) thus becomes

$$u_g^2 \approx \left(\frac{a_g}{a_p}\right)^2 (u_p^2 - V_p^2 F_p) + V_g^2 F_g \tag{6.59}$$

where

$$F_{p,g} \equiv \frac{\displaystyle\int_0^\infty f_{p,g}(r) E_p^2 r\, dr}{\displaystyle\int_0^\infty E_p^2 r\, dr} \tag{6.60}$$

In the above, $f_p(r) = (r/a_p)^2$. Using this, along with (6.56) in (6.60), it is found that $F_p = r_0^2/(2a_p^2)$. Finally, using (6.57) and (6.58), (6.59) becomes

$$u_g^2 \approx 2\left(\frac{a_g}{r_0}\right)^2 + V_g^2 F_g \tag{6.61}$$

The procedure is to find the graded index parameters, V_g and a_g, that will minimize (6.61).

As an example, consider the step index fiber, in which $V_g = V_s$ and $a_g = a_s$. Equation (6.60) with $f_g(r) = f_s(r)$ evaluates as $F_g = \exp(-2a_s^2/r_0^2)$. Equation (6.61) becomes

$$u_s^2 \approx 2\left(\frac{a_s}{r_0}\right)^2 + V_s^2 \exp\left(\frac{-2a_s^2}{r_0^2}\right) \tag{6.62}$$

Minimizing (6.62) is a matter of taking its derivative with respect to $(a_s/r_0)^2$ and setting the result to zero, obtaining:

$$r_0 = \frac{a_s}{\sqrt{\ln V_s}} \qquad (6.63)$$

and

$$u_s = [1 + \ln(V_s^2)]^{1/2} \qquad (6.64)$$

Performing the analysis for general alpha profile fibers requires the evaluation of F_g in (6.60) for the profile in question, followed by the minimization of u_g^2 in (6.61). The integrals involved are generally more complicated, and numerical solution methods must be used to obtain the results. Alternately, the wave equation can be numerically integrated directly to find LP_{01} mode parameters. Figure 6.12 shows r_0/a_g as a function of V_g for some representative alpha profiles as determined by the direct numerical integration method

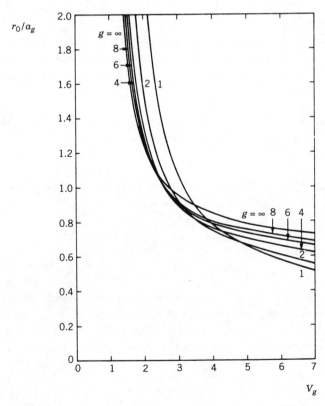

Figure 6.12. Normalized mode field radii, r_0/a_g, as functions of V_g for selected values of the profile parameter, α [16].

[16]. It is interesting to note that the r_0/a_g curves all cross within the region of V_g between 2 and 5, such that a decrease in α for a given V_g does not necessarily imply that the mode field radius will increase.

The curves of Fig. 6.12 are approximated by the following formula [16]:

$$\frac{r_0}{a_g} = \frac{A}{V_g^{2/(\alpha+2)}} + \frac{B}{V_g^{3/2}} + \frac{C}{V_g^6} \tag{6.65}$$

where

$$A = \left\{ \frac{2}{5} \left[1 + 4 \left(\frac{2}{\alpha} \right)^{5/6} \right] \right\}^{1/2} \tag{6.66}$$

$$B = \exp \left(\frac{0.298}{\alpha} \right) - 1 + 1.478[1 - \exp(-0.077\alpha)] \tag{6.67}$$

$$C = 3.76 + \exp \left(\frac{4.19}{\alpha^{0.418}} \right) \tag{6.68}$$

The use of a gaussian to approximate the fundamental mode is usually most accurate in the core region; minor inaccuracies occur in the cladding field, since the decay of the gaussian with radius generally occurs more rapidly than the actual evanescent field. Consequently, bending loss computations employing the gaussian can lead to results that are too low. On the other hand, propagation phenomena in the core and optimum coupling parameters are analyzed with acceptable accuracy using the gaussian.

6.5. ALTERNATIVE SINGLE-MODE INDEX PROFILES FOR CONTROL OF LOSS AND DISPERSION

A good single-mode fiber would exhibit the following characteristics: (1) low intrinsic loss, (2) low bending loss, (3) large mode field radius, thus allowing low coupling loss, and (4) zero or otherwise low dispersion at a specified wavelength or over a wavelength range of interest. It has been shown that the simple step index profile will not always enable all these characteristics to be realized at once. Grading the core index provides some added flexibility, specifically by enabling control of the mode field radius for a given Δ (although at a possible cost of increased bending loss); but this technique by itself is not sufficient to provide complete control of all the parameters. Other methods are used, which in their basic forms involve the use of more complicated step index structures. Such methods, when used in conjunction with index grading, provide the most flexibility in the design and have yielded the best results.

A primary issue in the design of a single-mode fiber is the choice of V number (or alternatively λ_c) for a given operating wavelength. It is V that determines the fraction of mode power that propagates in the cladding, as expressed in (3.85) and as shown in the curves of Fig. 3.11. It is best to minimize the cladding power fraction, since dopant losses (e.g., arising from the P_2O_5—OH resonance) and impurity absorption losses in the substrate portion of the cladding (when MCVD fabrication is used) can be appreciable. Thus it is best to keep V as high as possible, while maintaining a reasonable safety margin to prevent bimoded operation. This is equivalent to keeping λ_c as close as possible to λ. For fibers that operate at 1.3 μm, a successful practice has been to position λ_c between 1.0 and 1.2 μm; thus V will lie between 1.8 and 2.2.

Microbending loss must also be minimized for a given value of V. This loss will decrease with decreasing mode field radius, the latter being accomplished by decreasing the core radius. To keep V constant, Δ must be increased as the inverse square of a. Resulting is the requirement for a significant increase in the core index (assuming the cladding index is not changed). To accomplish this, the GeO$_2$ concentration in the core must be increased considerably, which results in the following two problems: (1) Rayleigh scattering loss in the core increases and (2) unwanted shifting of the zero dispersion wavelength, λ_0, may occur. The second effect will arise from both material and waveguide dispersion effects, but mostly from the former, provided the reduction of core size is relatively moderate.

The above problems were reduced to a satisfactory level by use of the *depressed cladding* index design, shown in Fig. 6.13a. In it, the cladding index within a region around the core is reduced; this enables the core index to be correspondingly lowered, thus allowing the required Δ to be achieved with less GeO$_2$ dopant. The *matched cladding* fiber having the same δn is shown in Fig. 6.13b. In the depressed cladding profile, the net Δ value is specified in terms of the core and cladding values, referenced to the outer cladding index; that is, $\Delta = \Delta^+ + \Delta^-$, where $\Delta^+ \equiv (n_1^2 - n_3^2)/2n_1^2$ and $\Delta^- \equiv (n_3^2 - n_2^2)/2n_3^2$. One stan-

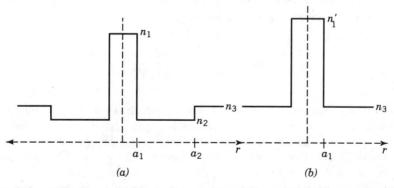

Figure 6.13. (a) Depressed cladding step index design and (b) matched cladding structure having the same δn.

dard design specifies the following values [17]: Δ^+ = 0.00255, Δ^- = 0.00115, $2a_1$ = 8.3 μm, and a_2/a_1 = 6.5.

As a rule, the outer radius, a_2, must be larger than about five times the core radius, to minimize the effect of the boundary at a_2 [18]. If this condition is not met, a substantial portion of the mode power will reside in the outer cladding, leading to the following effects: (1) material impurities and/or dopants in the outer cladding will contribute significantly to propagation losses and (2) the mode power may leak into the outer cladding since its index is higher than that of the inner cladding. The second effect leads to catastrophic loss for the mode and can occur when the operating wavelength—and hence the mode field radius—becomes too large. Thus, in depressed cladding fiber, an effective cutoff wavelength for LP_{01}, termed λ_f, may exist, such that operation is restricted for wavelengths between this value and the LP_{11} cutoff wavelength; that is, $\lambda_c < \lambda < \lambda_f$. While the depressed cladding design requires negligible power in the outer cladding, other applications make use of the outer cladding propagation effects for dispersion control purposes, as will be discussed later in this section. Such designs are usually referred to as *W profiles*.

In addition to minimizing losses, it is important to have control over the location of the zero dispersion wavelength. The ability to vary both core and cladding indices (such as in the depressed cladding design), as well as the core radius, provides enough parameters to allow simultaneous control of λ_0 (at wavelengths in the vicinity of 1.3 μm), while reducing losses to acceptable levels. Adding germania to silica shifts the dispersion to longer wavelengths, as shown in Fig. 5.6; adding fluorine to silica shifts the dispersion to shorter wavelengths. The combination of these two material dispersions, when added to the waveguide dispersion (in which the core radius plays a role), results in the ability to precisely control λ_0 between wavelengths of 1.28 and 1.38 μm with core radii between 3 and 5 μm [19].

In other applications, it is desired to shift the dispersion function to much longer wavelengths, so that, for example, λ_0 occurs at the minimum loss wavelength of 1.55 μm. Dopant-modified material dispersion does not provide enough shifting to reach 1.55 μm, but it can be done by increasing the amount of waveguide dispersion, given by (5.43):

$$D_w(\lambda) = -\frac{N_2^2 \Delta}{n_2 c \lambda} V \frac{d^2(bV)}{dV^2} \tag{6.69}$$

Increasing D_w is accomplished by raising Δ, which for a given V will again require a reduced core size, but to a much larger degree than was the case for the depressed cladding designs. Core radii on the order of 2 μm are typically required [20]. Consequently, much more scattering loss within the core is encountered in addition to losses thought to be associated with stress-induced defects in the vicinity of the core–cladding boundary [21]; the latter arise from the large index gradient at that location. These losses are higher than can adequately be compensated by a simple depressed cladding design. Furthermore, with the substantially reduced core size, the mode field radius becomes too small, likely resulting in higher coupling loss. These problems resulted in the exploration of graded index designs, which enable the reduction of boundary defects as the index gradient decreases; use of a graded core also results in the enlargement of the mode field radius for a given Δ. Grading functions that have been successful when used in dispersion-shifted designs include triangular ($\alpha = 1$) [20] and gaussian [8] profiles. Figure 6.14 shows some sample

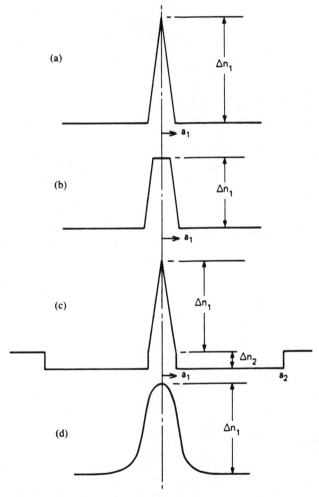

Figure 6.14. Index profile designs for dispersion-shifted fiber: (a) basic triangular core to reduce scattering losses, (b) trapezoidal core, (c) depressed cladding with triangular core (DDT fiber), and (d) gaussian profile [20]. © 1986 IEEE.

designs, including a combined triangular core/depressed cladding structure; the depressed cladding was used to reduce the mode field radius and thus decrease bending losses. With the latter design, a net loss of 0.24 dB/km at 1.55 μm was achieved, with λ_0 occurring at 1.54 μm [22].

In certain designs, it is possible to establish low dispersion over a broad range of wavelengths, such as between 1.3 and 1.6 μm. Such *dispersion-flattened* fibers typically have complicated profiles involving multiple steps, in addition to index grading to reduce losses. Most dispersion-flattening designs, however, are based on the relatively simple W profile fiber, mentioned earlier. A sketch of this profile is shown in Fig. 6.15.

The W profile consists of three regions: the *core* of index n_1 ($r < a_1$), the *inner cladding* having index n_2 ($a_1 < r < a_2$), and the *outer cladding* of index n_3 ($r > a_2$). The design is in fact recognized as a depressed cladding structure; it differs from the previously discussed cases, however, by typically having Δ^+ and Δ^- on the same order, and by having a_2 on the order of twice a_1 or less. As a result, the outer cladding plays a significant role in determining the waveguiding properties. As before, since $n_3 > n_2$, the possibility of leakage into the outer cladding exists, if the mode power in that region is appreciable.

To determine whether leakage of the mode can occur, consider the long-wavelength case (equivalent to $V \to 0$), in which the mode power distribution is essentially uniform over the fiber cross section. In this case, the mode encounters an effective core index, n_{avg}, that is found by averaging the refractive index over the area of cross section between $r = 0$ and $r = a_2$ [23]. The average cladding index ($r > a_2$) is n_3. Thus if $n_{avg} > n_3$, the mode will have a zero-frequency (infinite wavelength) cutoff, as in any conventional step index fiber. If $n_{avg} < n_3$, then cutoff for the mode will occur at some finite wave-

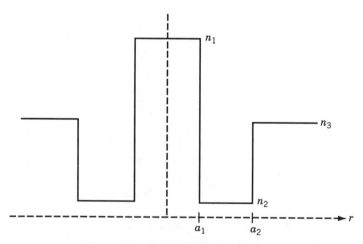

Figure 6.15. Basic W index profile design.

length, λ_f, associated with a nonzero normalized frequency, V_{co}. The average core index is found through

$$n_{avg} = \frac{1}{\pi a_2^2} \int_0^{2\pi} \int_0^{a_2} n(r) r \, dr \, d\phi \tag{6.70}$$

Using $n(r) = n_1$ for $r < a_1$ and $n(r) = n_2$ for $a_1 < r < a_2$, the integrations in (6.70) are performed to yield

$$n_{avg} = n_2 + (n_1 - n_2) \left(\frac{a_1}{a_2} \right)^2 \tag{6.71}$$

Using (6.71), the condition for a nonzero frequency cutoff ($n_3 > n_{avg}$) can be expressed as

$$\left(\frac{a_2}{a_1} \right)^2 > 1 + \frac{\Delta^+}{\Delta^-} \tag{6.72}$$

where $\Delta^+ \approx (n_1 - n_3)/n_3$ and $\Delta^- \approx (n_3 - n_2)/n_3$, assuming a weakly guiding structure.

To determine the values of V_{co} for a given design, a detailed study of the wave equation solutions in the three regions is used [23,24]. Specifically, the field solutions for a guided mode are assumed to be ordinary Bessel functions within the region $r < a_1$, a mixture of K and I modified Bessel functions within the region $a_1 < r < a_2$, and modified Bessel function, K, for $r > a_2$. Following analogous procedures to those used in Chapter 3, the tangential fields are matched at the two boundaries to yield an eigenvalue equation for the mode [23]. The cutoff condition for the mode is derived from the requirement that the fields exhibit no decay with radius in the outer cladding, resulting in a zero argument for the modified Bessel function in that region. The results are given by the curves of Fig. 6.16, showing V_{co} as functions of a_2/a_1 for selected values of Δ^-/Δ^+ [24]. In this analysis, V_{co} is defined as

$$V_{co} \equiv \frac{2\pi a_1}{\lambda_f} n_3 \sqrt{2\Delta^+} \tag{6.73}$$

Guided modes will thus occur for $V > V_{co}$, where the definition of V for this structure is, $V = (2\pi a_1/\lambda)\sqrt{n_1^2 - n_3^2}$. Note that the intercepts of the curves with the abscissa can be found from (6.72) and are given by $a_2/a_1 = \sqrt{1 + (\Delta^+/\Delta^-)}$.

How the W profile fiber flattens the dispersion can be understood by remembering that the mode field radius becomes larger with increasing wavelength,

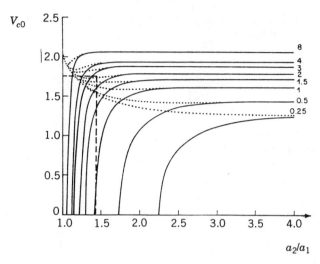

Figure 6.16. Curves of V_{c0} as functions of a_2/a_1 for selected values of Δ^-/Δ^+ [24]. © 1982 IEEE.

thus increasing the effect of the outer cladding. Additionally, the net group index of the fiber will depend on the percentages of the power that propagate in each region of the fiber cross section, as is determined for a step index fiber by (5.34). Thus a tightly confined mode will encounter a group index that is dominated by that of the core; on the other hand, a weakly confined mode will propagate with a lower group index that is close to that of the outer cladding. The transition that occurs between these group indices as wavelength (and mode size) changes is the key to the dispersion flattening principle.

Figure 6.17a shows a qualitative group index curve for a dispersion flattening W profile. Also shown are the group indices for the core and for the outer cladding. It is seen that, as wavelength increases, the net group index migrates from that of the core to that of the outer cladding (at which cutoff occurs), thus showing the effect of increasing the mode size. Resulting are two positions at which the net group index curve achieves a slope of zero, thus producing two zero dispersion wavelengths. Figure 6.17b shows the dispersion as calculated from the first derivative of the curve in Fig. 6.17a. The dual zero dispersion wavelength feature results in a partial flattening of the dispersion curve in the region between these two wavelengths. The primary design objective is thus to assure that the two zero crossings bracket the wavelength range of interest—typically between 1.3 and 1.6 μm. The zero location for the longer wavelength is a very sensitive function of the values of Δ^-/Δ^+ and a_2/a_1. For example, if a_2/a_1 is too large, then the outer cladding is too far away to have significant effect; on the other hand, too small a value of a_2/a_1 will cause the outer cladding to pull the group index down before the first dispersion zero is reached, so that the net dispersion monotonically decreases with wavelength,

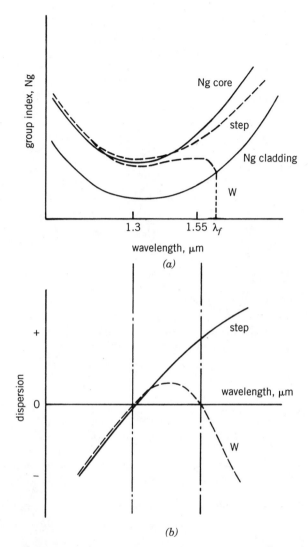

Figure 6.17. (a) Group index curves for a W profile fiber, showing the effects of the inner core and outer cladding. Solid curves: N for the core and cladding materials. Dashed curves: N for the W fiber and for a single-step index fiber. (b) Dispersion curves corresponding to the group index curves of (a). (Adapted from ref. 20.) © 1986 IEEE.

never achieving a zero. A detailed discussion of these issues is presented in ref. 20.

A problem with the basic W profile is that the presence of the outer cladding produces increased susceptibility to bending loss. This problem was reduced by incorporating an additional downward step in the outer cladding index at radius a_3, as shown in Fig. 6.18 [25]. The outer step serves to increase the confinement

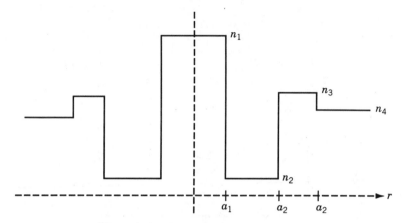

Figure 6.18. Quadruple-clad fiber structure.

of the mode such that bending losses are substantially reduced. A significant drawback of the modified structure is that the raised outer index ring provides good confinement of the LP_{11} and LP_{02} modes, whose intensity maxima occur in the ring [26]. Consequently, the outer step must be kept small, so that these modes will experience high loss and will thus carry a negligible amount of power.

PROBLEMS

6.1. Using Maxwell's equations, show that in a sourceless medium in which the permittivity varies with position, the wave equations for the electric and magnetic fields are expressed as (6.4) and (6.5).

6.2. Determine and sketch the form of Fig. 6.3 for the LP_{0m} mode case. Show that only one turning point exists and that, consequently, these modes are meridional.

6.3. Describe qualitatively the behavior of a given mode in a graded index fiber with frequency. Specifically, draw plots of $n^2(r)k_0^2 - l^2/r^2$ and β^2 as functions of radius showing the condition at cutoff and as frequency increases. Is there an increase or a decrease in turning point separation as frequency is raised? Use appropriate formulas to justify your answer.

6.4. In a graded index fiber with parabolic index profile ($\alpha = 2$), the LP_{01} mode field will assume the gaussian form:

$$E_x(r, z) = E_0 \exp[-(r/r_0)^2] \exp(-j\beta z)$$

Using the above field in the wave equation, show that the mode field radius, r_0, and the normalized radial propagation constant, u, will assume the forms $r_0 = a(2/V)^{1/2}$ and $u = (2V)^{1/2}$, so that

$$\beta = \left(\frac{1}{a}\right)\left(\frac{V^2}{2\Delta} - 2V\right)^{1/2}$$

In addition, show that the turning point radius of the mode as determined through the WKB method is equivalent to the mode field radius, as found above.

6.5. Consider the WKB method as applied to a step index fiber, having core radius a, core index n_1, and cladding index n_2. Light at radian frequency ω propagates in a single mode, having azimuthal mode number l and radial mode number m.

   a) Make a qualitative plot of $n^2(r)k_0^2 - l^2/r^2$ as a function of radius for the step index case. On the same plot indicate the function β^2, and label the turning points on the r axis.

   b) Determine the maximum value of β in terms of the given parameters.

6.6. Consider a multimode graded index fiber for which $\alpha = 1$ (triangular profile), having refractive index n_1 at $r = 0$, cladding index n_2, and core radius a. A mode for which $l = 0$ propagates in the fiber.

   a) Determine the approximate turning point locations at cutoff for the $l = 0$ mode.

   b) Derive an equation that gives the outer turning point location for the $l = 0$ mode as a function of the core radius and the normalized propagation constant, b, where the latter is defined in (3.78).

   c) Determine the range of possible values of b for this mode in this fiber.

6.7. In graded index fibers in which Δ is relatively high, the second-order term in the expression for group delay for the modes may not be negligible. Determine the optimum profile parameter, α_{opt}, for this case, such that the group delays for all propagating modes are equalized. Assume material dispersion is characterized by the parameter y.

6.8. Show that a graded index fiber link that exhibits optimum alpha performance can be constructed using two end-to-end coupled fibers whose values of α lie on either side of α_{opt}. Take $\alpha_1 = \alpha_{opt} + \delta_1$ and $\alpha_2 = \alpha_{opt} - \delta_2$, where δ_1 and δ_2 are both positive and much less than α_{opt}. Determine the relation between the fiber lengths and the δ values.

6.9. Light from a LED is to be coupled into an $\alpha = 2$ graded index fiber of core radius a, cladding index n_2, and on-axis core index n_1. The surface of the LED is circular, having radius r_s, where $r_s < a$.

a) Determine the numerical aperture angle of the fiber as a function of radius, $\theta_{N.A.}(r)$. Assume that the value of n_1 is close to that of n_2, so that $\sin \theta_{N.A.} \approx \theta_{N.A.}$.

b) The light from the LED is to be coupled into the fiber using a lens of magnification $M = 1$. The lens-to-fiber distance, L_2, is set to be as small as possible, while assuring that all power from the lens is coupled into the fiber. The LED emission obeys the Lambertian brightness function, $B = B_0 \cos \theta_s$. Determine the coupling efficiency, η, for this configuration, again assuming that all angles are small, such that $\sin \theta \approx \tan \theta \approx \theta$.

c) Determine the coupling efficiency for the case in which the diode is in coaxial contact with the front surface of the fiber. Compare this result to that of part b.

6.10. The gaussian refractive index profile in single-mode fibers has been shown to provide good overall performance and is readily fabricated using the VAD process. Consider the variational method employing the equivalent parabolic profile as applied to a gaussian profile specified by $f_g(r) = 1 - \exp[-(r/a_g)^2]$. Show that (6.51) becomes

$$ u_g^2 \approx 2 \left(\frac{a_g}{r_0} \right)^2 + V_g^2 \left[1 + 2 \left(\frac{a_g}{r_0} \right)^2 \right]^{-1} $$

and that minimizing u_g^2 in the above results in the mode field radius expression, $r_0 = \sqrt{2} a_g (V_g - 1)^{-1/2}$, and normalized radial propagation constant, $u_g = (2V_g - 1)^{1/2}$. Determine the value of V_g at which the mode field distribution is identical to that of the refractive index.

6.11. Using the alpha profile function (6.2), determine expressions for the guidance factor, G (6.54), and V_c (6.55) as functions of α.

REFERENCES

1. D. Gloge and E. A. J. Marcatili, "Multimode Theory of Graded-Core Fibers," *Bell System Technical Journal*, vol. 52, pp. 1563–1578, 1973.

2. R. Olshansky, "Multiple-α Index Profiles," *Applied Optics*, vol. 18, pp. 683–689, 1979.

3. S. Ramo, J. R. Whinnery, and T. Van Duzer, *Fields and Waves in Communication Electronics*, 2nd ed., John Wiley & Sons, New York, 1984, chap. 14.

4. R. Olshansky, "Propagation in Glass Optical Waveguides," *Reviews of Modern Physics*, vol. 51, pp. 341–367, 1979.

5. A. Cherin, *An Introduction to Optical Fibers*. McGraw-Hill, New York, 1983, chap. 6.

6. S. Geckeler, *Optical Fiber Transmission Systems*. Artech House, Norwood, MA, 1987.

7. R. Olshansky and D. B. Keck, "Pulse Broadening in Graded-Index Optical Fibers," *Applied Optics*, vol. 15, pp. 483–491, 1976.

8. R. Yamauchi, M. Miyamoto, T. Abiru, K. Nishide, T. Ohashi, O. Fukuda, and K. Inada, "Design and Performance of Gaussian-Profile Dispersion-Shifted Fibers Manufactured by VAD Process," *IEEE Journal of Lightwave Technology*, vol. LT-4, pp. 997–1004, 1986.

9. A. W. Snyder and R. A. Sammut, "Fundamental (HE_{11}) Modes of Graded Optical Fibers," *Journal of the Optical Society of America*, vol. 69, pp. 1663–1671, 1980.

10. H. Matsumura and T. Suganuma, "Normalization of Single-Mode Fibers Having an Arbitrary Index Profile," *Applied Optics*, vol. 19, pp. 3151–3158, 1980.

11. C. A. Millar, "Direct Method of Determining Equivalent Step-Index Profiles for Monomode Fibers," *Electronics Letters*, vol. 17, pp. 458–460, 1981.

12. P. M. Morse and H. Feshbach, *Methods of Theoretical Physics*. McGraw-Hill, New York, 1953, p. 1106H.

13. R. F. Harrington, *Time-Harmonic Electromagnetic Fields*. McGraw-Hill, New York, 1961, chap. 7.

14. W. A. Gambling, D. N. Payne, and H. Matsumura, "Cut-off Frequency in Radially Inhomogeneous Single-Mode Fibre," *Electronics Letters*, vol. 13, pp. 139–140, 1977.

15. W. A. Gambling, H. Matsumura, and C. M. Ragdale, "Wave Propagation in a Single-Mode Fibre with Dip in the Refractive Index," *Optical and Quantum Electronics*, vol. 10, pp. 301–309, 1978.

16. D. Marcuse, "Gaussian Approximation of the Fundamental Modes of Graded Index Fibers," *Journal of the Optical Society of America*, vol. 68, pp. 103–109, 1978.

17. S. R. Nagel, "Review of the Depressed Cladding Single-Mode Fiber Design and Performance for the SL Undersea System Application," *IEEE Journal of Lightwave Technology*, vol. LT-2, pp. 792–801, 1984.

18. B. J. Ainslee, K. J. Beales, C. R. Day, and J. D. Rush, "The Design and Fabrication of Monomode Optical Fiber," *IEEE Journal of Quantum Electronics*, vol. QE-18, pp. 514–523, 1982.

19. L. G. Cohen, W. L. Mammel, and S. Lumish, "Numerical Parametric Studies for Controlling the Wavelength of Minimum Dispersion in Germania Fluoro-phospho-silicate Single-Mode Fibres," *Electronics Letters*, vol. 18, pp. 38–39, 1982.

20. B. J. Ainslie and C. R. Day, "A Review of Single-Mode Fibers with Modified Dispersion Characteristics," *IEEE Journal of Lightwave Technology*, vol. LT-4, pp. 967–979, 1986 (and references cited therein).

21. B. J. Ainslie, K. J. Beales, C. R. Day, and J. D. Rush, "Interplay of Design Parameters and Fabrication Conditions on the Performance of Monomode Fibers Made by MCVD," *IEEE Journal of Quantum Electronics*, vol. QE-17, pp. 854–857, 1981.

22. H.-T. Shang, T. A. Lenahan, P. F. Glodis, and D. Kalish, "Design and Fabrication of Dispersion-Shifted Depressed-Clad Triangular-Profile (DDT) Single-Mode Fibre" *Electronics Letters*, vol. 21, pp. 201–203, 1985.

23. S. Kawakami and S. Nishida, "Characteristics of a Doubly Clad Optical Fiber with a Low-Index Inner Cladding," *IEEE Journal of Quantum Electronics*, vol. QE-10, pp. 879–887, 1974.

24. L. G. Cohen, D. Marcuse, and W. L. Mammel, "Radiating Leaky-Mode Losses in Single-Mode Lightguides with Depressed-Index Claddings," *IEEE Journal of Quantum Electronics*, vol. QE-18, pp. 1467–1472, 1982.

25. L. G. Cohen, W. L. Mammel, and S. J. Jang, "Low-Loss Quadruple-Clad Single-Mode Lightguides with Dispersion Below 2 ps/km-nm over the 1.28 μm to 1.65 μm Wavelength Range," *Electronics Letters*, vol. 18, pp. 1023–1024, 1982.

26. P. L Francois, F. Alard, J. F. Bayton, and B. Rose, "Multimode Nature of Quadruple-Clad Fibres," *Electronics Letters*, vol. 20, pp. 37–38, 1984.

7

Outline of Nonlinear Optics

The previous chapters have described the propagation of light through optical fibers under the assumption that the material permittivity exhibits variation only with position and frequency. The frequency dependence was found to arise from resonances that were centered at frequencies far from those of interest. As a result of this condition, along with the assumption of low field intensities, the classical picture demonstrates that electrons or atoms that are set into oscillation under the influence of a time-varying field will do so at relatively low amplitude and as simple harmonic oscillators. The phase difference between the light reradiated by the dipole oscillators and the incident light produces a net phase velocity for the total field that is different from that of the incident field—an effect that is accounted for in the wave equation by introducing the frequency-dependent refractive index.

In this chapter, the restriction of low field intensity is removed, thus allowing a number of interesting material effects that were previously negligible to have a significant influence on the propagating field. The magnitudes of these new effects exhibit a nonlinear dependence on the incident field strength. They are thus known as *nonlinear optical phenomena*. The field of nonlinear optics is devoted to the study of these effects—chiefly with regard to understanding the conditions necessary for a given nonlinear process to occur, the identification of materials that will either enhance or prevent the process, and applications, of which there are many. The nonlinear processes that can occur in materials are often complicated, and the various effects can be interrelated. Multiple effects can occur simultaneously in a given situation, and a single effect can be produced simultaneously by more than one mechanism.

As an example, consider a material composed of molecules that exhibit vibrational or rotational resonances of relatively low frequency compared to that of the incident optical field. Intense light incident on the substance may excite the molecules into oscillation *at their resonant frequencies*. The oscillation directly affects the refractive index encountered by the light, in that the value of n will be slightly modulated at the resonant frequency. The light will thus "beat" with the time-varying index to produce additional light that is up- or down-shifted in frequency. The generated light can then further interact with the input light, producing a difference frequency that is tuned to the material resonance. Thus more molecules are excited and more conversion takes place from one light frequency to the other. This description of the effect known as stimulated Raman scattering is greatly oversimplified here. Nevertheless, it provides a feel for the processes that are encountered.

Optical fibers are capable of supporting a number of nonlinear effects, despite the fact that the strengths of the nonlinearities in fused silica are as a rule relatively low compared to many other materials. It will be seen that one of the requirements to achieve a large nonlinear signal of any type is that there be sufficient *interaction length* in the material over which the process can occur, thus allowing the signal to grow to appreciable levels. In view of this, any deficiency in the strenth of a nonlinear mechanism in a fiber is usually compensated by the extremely long lengths available. An additional factor that enhances fiber nonlinearities is the small core area; this results in mode field strengths being high, even though the total power carried by the fiber is relatively modest. The different nonlinear processes in a fiber can be either useful or undesirable, depending on the application. In modern communication systems that involve high power levels and that employ optical amplifiers, the understanding and utility of nonlinear effects have become extremely important.

7.1. ROLE OF MEDIUM POLARIZATION IN WAVE PROPAGATION

In linear optics, the macroscopic medium polarization at a specified frequency and position in space is related to the local electric field through the linear susceptibilty:

$$\mathscr{P} = \epsilon_0 \chi_L \mathscr{E} \qquad (7.1)$$

The electric flux density, $\mathscr{D}$, is defined in terms of the electric field and polarization by

$$\mathscr{D} = \epsilon_0 \mathscr{E} + \mathscr{P} \qquad (7.2)$$

Using (7.1) and (7.2) in Maxwell's equations in a manner similar to that shown in Chapter 1, the wave equation for the electric field becomes

$$\nabla^2 \mathscr{E} - \mu_0 \epsilon_0 \frac{\partial^2 \mathscr{E}}{\partial t^2} = \mu_0 \frac{\partial^2 \mathscr{P}}{\partial t^2} \qquad (7.3)$$

where the medium is assumed sourceless and nonmagnetic. The dielectric constant is defined in terms of the linear susceptibility through $\epsilon_r = 1 + \chi_L$. Using this, (7.3) is condensed into the more familiar form

$$\nabla^2 \mathscr{E} - \mu_0 \epsilon_r \epsilon_0 \frac{\partial^2 \mathscr{E}}{\partial t^2} = 0 \tag{7.4}$$

Equation (7.3) indicates that the time-varying polarization serves as a driving term in the wave equation. Physically, the atomic dipoles that make up the macroscopic polarization have been excited into oscillation by the incident electric field. These in turn reradiate to produce "secondary" waves that add to the original field in some phase relationship; the phase depends on the position of the incident field frequency relative to the resonant frequency of the dipole oscillators. The superposition of the two fields at each point in the medium results in a retardation or advance in the net phase compared to that of the incident field. This effect is manifested as an observed increase or decrease in the phase velocity of the wave in the material. The phase shift phenomenon is cumulative over the length of the medium, since the primary field at a given position in fact includes the incident field as well as all secondary fields that were radiated from positions further back. This total field is represented as $\mathscr{E}$ in (7.3) and (7.4). The magnitude of the net phase shift between light that has propagated through a block of the material and light that traveled over an equal distance in free space will increase linearly with the medium length. The length dependence on the magnitude of the effect (in this case phase shift), while being obvious for the case of linear optics, is also a very important principle in nonlinear optics.

7.2. NONLINEAR POLARIZATION

To account for nonlinear effects in general media, the macroscopic polarization is written in the form of a power series expansion in the electric field strength:

$$\mathscr{P} = \epsilon_0 [\chi^{(1)} \mathscr{E} + \chi^{(2)} \mathscr{E}\mathscr{E} + \chi^{(3)} \mathscr{E}\mathscr{E}\mathscr{E} + \cdots] = \mathscr{P}_L + \mathscr{P}_{NL} \tag{7.5}$$

where $\chi^{(1)} = \chi_L$ is a second rank tensor describing linear polarization (allowing for the possibility of anisotropic media). The linear polarization, $\mathscr{P}_L$, is given by (7.1). Terms of second and higher order in (7.5) are collectively referred to as the nonlinear polarization, $\mathscr{P}_{NL}$. Most work in nonlinear optics involves second or third order processes, characterized by the associated susceptibilities, $\chi^{(2)}$ and $\chi^{(3)}$. Higher order processes are as a rule extremely weak and have not been generally applicable to practical use. Virtually all effects in optical fibers are third order, although effects that mimic second order processes have been observed.

The products of vector fields appearing in (7.5) indicate products between all spatial field components. Thus the product $\mathscr{E}\mathscr{E}$ would in general contain 9 terms and $\mathscr{E}\mathscr{E}\mathscr{E}$ would expand to 27 terms. The susceptibilities, $\chi^{(2)}$ and

$\chi^{(3)}$, are thus third and fourth rank tensors that could contain 27 and 81 different terms, respectively. This is in fact rarely the case, since most materials exhibit symmetries that significantly reduce the number of independent terms. In isotropic media, such as straight optical fibers of circular cross section, and in which all light wavelengths are far from any resonance, the number of independent terms in the third order susceptibility is reduced to only one.

The effect on the wave equation is to introduce a new driving term in (7.4), which accounts for the nonlinear polarization:

$$\nabla^2 \mathscr{E} - \mu_0 \epsilon_r \epsilon_0 \frac{\partial^2 \mathscr{E}}{\partial t^2} = \mu_0 \frac{\partial^2 \mathscr{P}_{NL}}{\partial t^2} \tag{7.6}$$

where the linear polarization has been absorbed into ϵ_r, which in general is a tensor. The nonlinear polarization accounts for the radiation of light at frequencies that may differ from those of the incident waves, and which may propagate in different directions. The electric field, $\mathscr{E}$, in (7.5) and (7.6) represents the *total* field—that is, the sum of all incident fields and those generated by the nonlinear polarization. In cartesian coordinates, the vector form of $\mathscr{E}$ would be

$$\mathscr{E} = \mathscr{E}_x \hat{\mathbf{a}}_x + \mathscr{E}_y \hat{\mathbf{a}}_y + \mathscr{E}_z \hat{\mathbf{a}}_z \tag{7.7}$$

Each component of $\mathscr{E}$ is in turn represented as a sum over all frequencies and propagation directions present. For example, the x component of $\mathscr{E}$ would be

$$\mathscr{E}_x = \sum_m \sum_n \tfrac{1}{2} E_{x0}^{mn} \exp(j\omega_m t) \exp(-j\mathbf{k}_n \cdot \mathbf{r}) + \text{c.c.} \tag{7.8}$$

The complex conjugate (c.c.) terms are included since all interacting fields are real, and many important nonlinear polarization terms would be missing without them. In linear optics, it is permissible to omit the complex conjugate (as was done in earlier chapters) since its only effect in the analysis is to produce a separate redundant wave equation. It should be noted that the individual susceptibility tensor elements apply to products of field vector components, such as $\mathscr{E}_x \mathscr{E}_y$, and so on. Since each component may contain a number of terms at different frequencies, as shown in (7.8), the tensor element must in general be reevaluated for each frequency present in the field product expansion. This procedure, although very tedious, often yields relatively simple results, since generally most of the terms in such an evaluation have negligible effect on the final nonlinear polarization for various reasons. One who is experienced in dealing with such calculations is usually able to immediately select the few important terms.

Once the nonlinear polarization is determined, it, along with the total electric field, is substituted into the wave equation. The latter is usually found to separate into a set of coupled equations that describe the growth or decay of a field of interest as functions of time and distance through the medium. An example of such a calculation is given in a later section.

7.3. STRUCTURE OF THE NONLINEAR SUSCEPTIBILITY

Nonlinear effects can arise from a host of mechanisms that depend on the type of material and its composition. It should be noted that more than one process can be occurring simultaneously within a given material for a certain input wave configuration, or a single nonlinear process can arise from more than one mechanism simultaneously.

A common classical model used to describe second order processes is the *anharmonic oscillator*. Consider the spring electron oscillator model discussed in Chapter 1. In the anharmonic case, the restoring force on the electron exhibits a dependence on the square of its displacement in a single coordinate direction, or as a product of displacements along two orthogonal directions. Physically, this represents a departure from the usual assumption of a linear restoring force on the electron. The linear assumption breaks down at large displacements, at which forces from neighboring atoms begin to be important. This effect is exemplified in Fig. 7.1, in which the restoring forces on an electron in a molecule are accounted for in three spring constants—each of the three is associated with a single coordinate direction. The individual spring constants are also nonlinear with displacement. It can be seen that small electron displacement along one direction would result in negligible stretching of the other two springs. On the other hand, large displacement could result in nonlinear

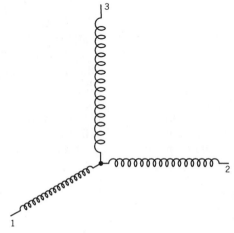

Figure 7.1. Three-dimensional spring model used in the nonlinear susceptibility derivation.

behavior of the parallel spring and could result in additional forces produced by the two perpendicular springs.

With electric field at frequency ω applied along the i coordinate direction, the resulting equation of motion that accounts for the above effects is

$$\frac{d^2q_i}{dt^2} + \zeta_i\,\frac{dq_i}{dt} + \omega_{0i}^2 q_i = \frac{-e\mathscr{E}_i^{(\omega)}}{m} - \sum_{j=1}^{3}\sum_{k=1}^{3}\frac{k_{ijk}}{m}\,q_j q_k \qquad (7.9)$$

where $q_{i,j,k}$ is the electron displacement from equilibrium in any of the three coordinate directions, ζ_i is the damping coefficient associated with motion in the i direction, and k_{ijk} is the strength of coupling between the subscripted displacements. The latter two quantities are treated phenomenologically in this discussion. Their values can be determined either through measurement or by quantum theory calculations.

Consider the special case of second harmonic generation, in which a nonlinear polarization is obtained that oscillates at frequency 2ω. To solve (7.9), a series solution for the displacement is used, in which each successive term oscillates at twice the frequency of the previous one. Using the first two terms of such a sequence (oscillating at frequencies ω and 2ω), it is left as an exercise to show that the resulting second order susceptibility associated with the second harmonic polarization is

$$\chi_{ijk}^{(2)}(2\omega;\omega,\omega)$$

$$= \frac{Ne^3 k_{ijk}}{\epsilon_0 m^3[(\omega_{0i}^2 - 4\omega^2) + j2\omega\zeta_i][(\omega_{0j}^2 - \omega^2) + j\omega\zeta_j][(\omega_{0k}^2 - \omega^2) + j\omega\zeta_k]}$$

$$(7.10)$$

where $\omega_{i,j,k}$ and $\zeta_{i,j,k}$ are the resonant frequencies and damping coefficients for electron oscillation in the three coordinate directions. The argument of the susceptibility $(2\omega;\omega,\omega)$ specifies that a nonlinear polarization that oscillates at frequency 2ω will be established as a result of two inputs, each at frequency ω.

Third order processes can be described by an anharmonic oscillator model having a restoring force term proportional to the product of three displacements. It is assumed that second order effects do not occur. The electron displacement equation becomes

$$\frac{d^2q_i}{dt^2} + \zeta_i\,\frac{dq_i}{dt} + \omega_{0i}^2 q_i = \frac{-e\mathscr{E}_i^{(\omega)}}{m} - \sum_{j=1}^{3}\sum_{k=1}^{3}\sum_{l=1}^{3}\frac{k_{ijkl}}{m}\,q_j q_k q_l \qquad (7.11)$$

The solution of (7.11) produces the third order susceptibility, one element of which is shown below for the general case of four different frequencies:

$$\chi_{ijkl}^{(3)}(\omega_d; \omega_a, \omega_b, \omega_c) =$$

$$\frac{Ne^4 k_{ijkl}}{4\epsilon_0 m^4 [(\omega_{0i}^2 - \omega_d^2) + j\omega_d \zeta_i][(\omega_{0j}^2 - \omega_a^2) + j\omega_a \zeta_j][(\omega_{0k}^2 - \omega_b^2) + j\omega_b \zeta_k][(\omega_{0l}^2 - \omega_c^2) + j\omega_c \zeta_l]}$$

$$(7.12)$$

It is seen that the forms of the second and third order susceptibilities as depicted in (7.10) and (7.12) exhibit similar resonance structures. $\chi^{(3)}$ in this case is oversimplified, however, since the spring model accounts only for single-photon resonances. The other possibilities would include two-photon and Raman effects, which would incorporate additional resonant denominators involving sum and difference frequencies of the form $[\omega_{0t}^2 - (\omega_c + \omega_b)^2 + j(\omega_c + \omega_b)\zeta_t]$, for a two-photon process, and $[\omega_{0r}^2 - (\omega_c - \omega_b)^2 + j(\omega_c - \omega_b)\zeta_r]$, for a Raman process.

Since the second order nonlinear polarization involves the product of two fields, an important symmetry requirement arises for media in which such processes are to occur. Consider, for example, the special case of copolarized input fields. The resulting nonlinear polarization will assume an orientation determined by the form of the susceptibility tensor. Now consider "inverting" the medium across a plane perpendicular to the applied field direction. This has the effect of replacing the unit vector $\hat{a}_f$, which specifies the field direction, with $-\hat{a}_f$—in other words, the same effect can be obtained just by reversing the applied field directions. If the medium is symmetric with respect to this inversion, the susceptibility should not change, and it should logically follow that the direction of the nonlinear polarization should reverse along with the fields. But since the polarization is proportional to the product of the two fields, such a reversal would not be allowed mathematically. It therefore follows that the tensor element corresponding to this interaction geometry must be zero, as will all such elements associated with input geometries to which inversion symmetry applies. If, on the other hand, the material is nonsymmetric with respect to inversion for a certain direction, then it becomes possible to achieve a single nonlinear polarization direction whether the input field direction is reversed or not. Recall that the nonlinear polarization represents a small perturbation on the actual electron displacement, and that the *overall* electron displacement will in fact reverse as the input fields change direction. The magnitude of the overall displacement, however, may be different for the two directions, or the electron trajectory may assume a curved path that could give rise to a second order polarization in a different direction and that satisfies the sign reversal requirements of $\mathscr{P}_{NL}^{(2)}$. A few good pictorial examples are given in [1].

The requirement of non-centro-symmetric media can only be satisfied by crystalline materials that lack inversion symmetry. Crystal lattices alone are entirely symmetric with respect to inversion; but with the inclusion of the material molecular structure at each lattice point, inversion symmetry is in general lost in certain directions. Amorphous materials, such as fused silica used in the production of optical fibers, are thus ruled out as medium candidates for second order effects, since they are by nature completely symmetric. Only third or higher order effects of odd order are possible in amorphous media, since the product of an odd number of fields appearing in the nonlinear polarization expression produces the required sign change in $\mathscr{P}_{NL}^{(3)}$ when the symmetric medium is inverted. The remainder of this chapter will be concerned with third order effects.

7.4. SYMMETRIES IN THE THIRD ORDER SUSCEPTIBILITY TENSOR

The third order susceptibility is represented by $\chi^{(3)}$, which is a fourth rank tensor, having 81 elements. Fortunately, most media possess symmetries that render many tensor elements zero or cause many to be equal. For a thorough discussion of the form of $\chi^{(3)}$ in various classes of media, the reader is referred to ref. 2.

Isotropic materials, from which optical fibers are constructed, have by definition complete spatial symmetry; as a result, the number of independent tensor elements is reduced to only three for near-resonant operation or just one for operation off-resonance. Consider, for example, the interaction diagrammed in Fig. 7.2, in which a nonlinear polarization component along the 3 axis is to be generated by three input field components along the 1, 2, and 3 axes, all of which have different frequencies. The polarization of interest is of frequency $\omega_d = \omega_a + \omega_b + \omega_c$. The fields are expressed in the form

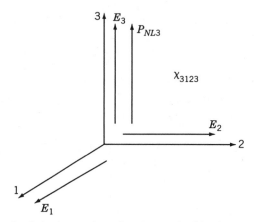

Figure 7.2. Field and polarization configuration characterized by the susceptibility tensor element χ_{3123}.

$$\mathscr{E}_i = \tfrac{1}{2} E_{0i} \exp(-j\mathbf{k}_i \cdot \mathbf{r}) \exp(j\omega_u t) + \text{c.c.}$$
$$= \tfrac{1}{2} E_i \exp(j\omega_u t) + \text{c.c.} \qquad i = 1, 2, 3, \quad u = a, \dots, d \qquad (7.13)$$

where E_i is the phasor form of the electric field polarized in the i direction. The nonlinear polarization is expressed in a similar way:

$$\mathscr{P}_{NL_i} = \tfrac{1}{2} P_{NL_i} \exp(j\omega_d t) + \text{c.c.} \qquad (7.14)$$

in which P_{NL_i} is the nonlinear polarization phasor. The total field is now

$$\mathscr{E} = \sum_{i=1}^{3} \mathscr{E}_i \hat{\mathbf{a}}_i \qquad (7.15)$$

The nonlinear polarization is then expressed as

$$\mathscr{P} = \epsilon_0 \chi^{(3)} \mathscr{E}\mathscr{E}\mathscr{E} \qquad (7.16)$$

By using (7.13) through (7.16), the polarization phasor for the interaction of interest is found in terms of the field phasors through

$$P_{NL_3} = \tfrac{6}{4} \chi_{3123}(\omega_d; \omega_a, \omega_b, \omega_c) E_1(\omega_a) E_2(\omega_b) E_3(\omega_c) \qquad (7.17)$$

The factor of 6 arises from the fact that this particular product of fields occurs 6 times in the computation of (7.16). Other field products in (7.16) may have different degeneracy factors. The reader is encouraged to write out (7.16) explicitly using (7.13) and (7.15) and note that degeneracies of 1, 3, or 6 will occur for the different terms.

The tensor element χ_{3123} is zero if the medium is isotropic. This is understood by observing the effect on the polarization by inverting one of the axes that is perpendicular to the polarization—the 2 axis, for example. This has the effect of replacing the 2 axis coordinates with their negative counterparts, thus changing the sign of E_2 appearing in (7.17) and, correspondingly, the sign of P_{NL_3}. Physically, the medium has not changed since it is isotropic, and so the sign change in the polarization would make no sense. Therefore it must follow that in isotropic media, $\chi_{3123} = 0$ and so must all other tensor elements possessing indices that are not paired. This important condition renders zero 60 of the 81 elements in an isotropic material. The remaining 21 are simply related since any interaction should be unchanged by simply rotating the medium. The requirement of invariance as a result of rotating the medium through 90 degrees results in the following equalities:

$$\chi_{1111} = \chi_{2222} = \chi_{3333} \tag{7.18}$$

$$\chi_{1122} = \chi_{2211} = \chi_{3311} = \chi_{1133} = \chi_{2233} = \chi_{3322} = \chi_{(1)} \tag{7.19}$$

$$\chi_{1212} = \chi_{2121} = \chi_{3131} = \chi_{1313} = \chi_{2323} = \chi_{3232} = \chi_{(2)} \tag{7.20}$$

$$\chi_{1221} = \chi_{2112} = \chi_{3113} = \chi_{1331} = \chi_{2332} = \chi_{3223} = \chi_{(3)} \tag{7.21}$$

The requirement of invariance through any angle of rotation results in the following relation, the proof of which is left as an exercise:

$$\chi_{1111} = \chi_{(1)} + \chi_{(2)} + \chi_{(3)} \tag{7.22}$$

If, in addition, all frequencies are far off resonance, the *Kleinman symmetry* condition holds, which enables all indices within each tensor element to be permuted freely without changing the value of the element. Thus off resonance,

$$\chi_{(1)} = \chi_{(2)} = \chi_{(3)} = \frac{\chi_{1111}}{3} \tag{7.23}$$

Hence an isotropic transparent material possesses only one *independent* third order susceptibility tensor element.

7.5. EXAMPLE: THIRD HARMONIC GENERATION

To demonstrate the manner in which the nonlinear polarization is used in the wave equation, an idealized example is presented here involving third harmonic generation. The wave geometry is shown in Fig. 7.3. Two input waves having crossed polarizations propagate collinearly in the $+z$ direction within a lossless isotropic medium in which third order processes can occur. Both waves have the same frequency, ω; their interaction via the third order susceptibility

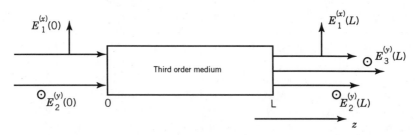

Figure 7.3. Beam geometry for the third harmonic generation example.

results in a third wave at frequency 3ω. Third harmonic waves with both x and y polarizations would in general be produced. The wave of interest in this case is that having y polarization. Additionally, the field envelopes are assumed not to vary with time. The two input fields are expressed in real form as

$$
\begin{aligned}
\mathscr{E}_1 &= \tfrac{1}{2} E_{01} \exp(-jk^\omega z) \exp(j\omega t) \hat{\mathbf{a}}_x + \text{c.c.} \\
&= \tfrac{1}{2} E_1 \exp(j\omega t) \hat{\mathbf{a}}_x + \text{c.c.} \tag{7.24} \\
\mathscr{E}_2 &= \tfrac{1}{2} E_{02} \exp(-jk^\omega z) \exp(j\omega t) \hat{\mathbf{a}}_y + \text{c.c.} \\
&= \tfrac{1}{2} E_2 \exp(j\omega t) \hat{\mathbf{a}}_y + \text{c.c.} \tag{7.25}
\end{aligned}
$$

The generated field of interest is

$$
\begin{aligned}
\mathscr{E}_3 &= \tfrac{1}{2} E_{03} \exp(-jk^{3\omega} z) \exp(j3\omega t) \hat{\mathbf{a}}_y + \text{c.c.} \\
&= \tfrac{1}{2} E_3 \exp(j3\omega t) \hat{\mathbf{a}}_y + \text{c.c.} \tag{7.26}
\end{aligned}
$$

where the propagation constants are $k^\omega = n^\omega \omega/c$ and $k^{3\omega} = 3n^{3\omega}\omega/c$. Note that the use of the subscripts, 1, 2, and 3 in this example does not refer to the coordinate axes, but enables the fields to be distinguished from one another.

The total field is now $\mathscr{E}_T = \mathscr{E}_1 + \mathscr{E}_2 + \mathscr{E}_3$, and the nonlinear polarization is $\mathscr{P}_{NL} = \epsilon_0 \chi^{(3)} \mathscr{E}_T^3$. This expansion is carried out under the assumption that the amplitudes E_{02} and E_{03} are much less than E_{01}. Thus only terms involving products of E_{01} with either E_{02} or E_{03} are retained. These give rise to nonlinear polarizations that oscillate at frequencies ω and 3ω. The resulting y-directed polarization amplitudes are

$$
\begin{aligned}
P^\omega_{NL_y} &= \tfrac{3}{4} \epsilon_0 \chi_{yxxy}(\omega; \omega, -\omega, \omega)[2|E_{01}|^2 E_{02} + E_{01}^2 E_{02}^*] \exp(-jk^\omega z) \\
&\quad + \tfrac{3}{4} \epsilon_0 \chi_{yxxy}(\omega; -\omega, -\omega, 3\omega)(E_{01}^*)^2 E_{03} \\
&\quad \cdot \exp[-j(k^{3\omega} - 2k^\omega)z] \tag{7.27} \\
P^{3\omega}_{NL_y} &= \tfrac{6}{4} \epsilon_0 \chi_{yxxy}(3\omega; \omega, -\omega, 3\omega)|E_{01}|^2 E_{03} \exp(-jk^{3\omega} z) \\
&\quad + \tfrac{3}{4} \epsilon_0 \chi_{yxxy}(3\omega; \omega, \omega, \omega)E_{01}^2 E_{02} \exp(-j3k^\omega z) \tag{7.28}
\end{aligned}
$$

The net nonlinear polarization that appears in the wave equation (7.6) is now $\mathscr{P}_{NL} = \mathscr{P}^\omega_{NL} + \mathscr{P}^{3\omega}_{NL}$, where the two right-hand side terms are constructed from the amplitudes in (7.27) and (7.28) using (7.14). Since all fields are plane waves, spatial variation will occur only in the z direction. Equation (7.6) thus assumes the form

$$\frac{\partial^2 \mathscr{E}_T}{\partial z^2} - \mu_0 \epsilon_0 n^2 \frac{\partial^2 \mathscr{E}_T}{\partial t^2} = \mu_0 \frac{\partial^2 \mathscr{P}_{NL}}{\partial t^2} \tag{7.29}$$

where n, the linear refractive index, is frequency dependent.

Now, when substituting the total field into the first term in (7.29), the result for each term is

$$\frac{\partial^2 \mathscr{E}_i}{\partial z^2} = \frac{1}{2} \left(\frac{\partial^2 E_{0i}}{\partial z^2} - j2k \frac{\partial E_{0i}}{\partial z} - k^2 E_{0i} \right) \exp[j(\omega t - kz)] + \text{c.c.}$$

$$i = 1, 2, 3, \quad k = k^\omega, k^{3\omega} \tag{7.30}$$

At this stage two additional assumptions are made. First, it is assumed that the x-polarized input field is strong enough such that it experiences negligible change in propagating through the medium. In other words, a very small percentage of its power is converted into either of the two other waves. Therefore we obtain $\partial^2 E_{01}/\partial z^2 = \partial E_{01}/\partial z = 0$. The second assumption is that the weaker fields, E_2 and E_3, will exhibit slow variation over the medium length, such that all second derivatives with respect to z are negligible. This condition is formally expressed as the *slowly varying envelope approximation* (SVEA), which is

$$\left| k \frac{\partial E_{0i}}{\partial z} \right| \gg \left| \frac{\partial^2 E_{0i}}{\partial z^2} \right| \tag{7.31}$$

Since for a plane wave $k = 2\pi/\lambda$, the above can be rewritten as

$$\left| \lambda \frac{\partial}{\partial z} \left(\frac{\partial E_{0i}}{\partial z} \right) \right| \ll \left| \frac{\partial E_{0i}}{\partial z} \right| \tag{7.32}$$

This can be interpreted as stating that the change in slope of the field envelope over a distance of one wavelength is much less than the magnitude of the slope itself. This approximation is valid for nearly all types of inputs, pulsed or otherwise, but loses validity for pulses of durations on the order of a few optical cycles.

With the above assumptions, the fields and polarizations are substituted into (7.29) to obtain

$$\left((k^\omega)^2 (E_{01} + E_{02}) + j2k^\omega \frac{\partial E_{02}}{\partial z} \right) \exp\left[j(\omega t - k^\omega z) \right]$$

$$+ \left((k^{3\omega})^2 E_{03} + j2k^{3\omega} \frac{\partial E_{03}}{\partial z} \right) \exp\left[j(3\omega t - k^{3\omega} z) \right]$$

$$- \mu_0 \epsilon_0 (n^\omega)^2 \omega^2 (E_{01} + E_{02}) \exp\left[j(\omega t - k^\omega z) \right]$$

$$- \mu_0 \epsilon_0 (n^{3\omega})^2 (3\omega)^2 E_{03} \exp\left[j(3\omega t - k^{3\omega} z) \right]$$

$$= \mu_0 \omega^2 P^\omega_{NL_y} \exp(j\omega t) + \mu_0 (3\omega)^2 P^{3\omega}_{NL_y} \exp(j3\omega t) \tag{7.33}$$

where $P^\omega_{NL_y}$ and $P^{3\omega}_{NL_y}$ are given in (7.27) and (7.28). A separate redundant equation, analogous to (7.33), results from the conjugate terms in the total field and polarization.

Now, noting that $k^\omega = n^\omega \omega/c$ and $k^{3\omega} = 3n^{3\omega}\omega/c$, a number of terms in (7.33) cancel. The equation separates into two by equating terms in ω and in 3ω. Resulting are two coupled first order differential equations describing the growth of E_{02} and E_{03} with distance in the medium:

$$\frac{dE_{02}}{dz} = \frac{-j3\omega}{8n^\omega c} \{ \chi_{yxxy}(\omega; \omega, -\omega, \omega)[2|E_{01}|^2 E_{02} + E_{01}^2 E_{02}^*]$$

$$+ \chi_{yxxy}(\omega; -\omega, -\omega, 3\omega) E_{01}^{*2} E_{03} e^{j\Delta k z} \} \tag{7.34}$$

$$\frac{dE_{03}}{dz} = \frac{-j3\omega}{8n^{3\omega} c} \{ 6\chi_{yxxy}(3\omega; \omega, -\omega, 3\omega)|E_{01}|^2 E_{03}$$

$$+ 3\chi_{yxxy}(3\omega; \omega, \omega, \omega) E_{01}^2 E_{02} e^{-j\Delta k z} \} \tag{7.35}$$

where the *phase mismatch* per unit distance, Δk, is

$$\Delta k = 3k^\omega - k^{3\omega} \tag{7.36}$$

Consider the special case of weak coupling, such that $E_{03} \ll E_{02}$ and that the change in E_{02} is negligible; that is, $dE_{02}/dz \approx 0$. In this case, (7.35) becomes

$$\frac{dE_{03}}{dz} = \frac{-j9\omega}{8n^{3\omega} c} \chi_{yxxy}(3\omega; \omega, \omega, \omega) E_{01}^2 E_{02} e^{-j\Delta k z} \tag{7.37}$$

The above can be directly integrated over z from 0 to L to obtain

$$E_{03}(L) = \frac{-j9\omega \chi_{yxxy}}{8n^{3\omega} c} E_{01}^2 E_{02} L \frac{\sin(\Delta k L/2)}{\Delta k L/2} \exp(-j\Delta k L/2) \tag{7.38}$$

where $E_{03}(0) = 0$. It is seen that E_{03} maximizes when $\Delta k = 0$, in which case the interaction is *phase matched*. In the present example in which all beams are collinear, (7.36) indicates that phase matching is possible only if the refractive indices at the two wavelengths are the same—a situation not possible in isotropic materials.

7.6. NONLINEAR REFRACTIVE INDEX

In the last section, third harmonic generation was treated by solving the wave equation (7.29) using the nonlinear polarization in (7.28), which oscillates at 3ω. A by-product of this interaction is embodied in the nonlinear polarization terms at the original frequency, ω, given in (7.27). Of particular interest are the first two terms in (7.27), which arise solely from inputs at ω. These polarization terms radiate waves at ω that are superimposed on the original input waves. The phase difference between input and radiated fields increases with length through the medium. The net result is that the net field propagates at a slower or a faster velocity through the material, depending on the direction of the phase shift. This effect is embodied in the refractive index, which is now observed to have a nonlinear component whose strength is dependent on the magnitude of the nonlinear polarization.

A simplified example of this is that in which the three input fields are copolarized. In this case a nonlinear polarization term similar to that given by the first term in (7.27) is obtained, but which involves χ_{xxxx}. The phasor polarization is

$$P^{\omega}_{NL_x} = \tfrac{3}{4}\,\epsilon_0 \chi_{xxxx}(\omega;\omega,-\omega,\omega)|E_{01}|^2 E_{01}\exp(-jk^{\omega}z) \qquad (7.39)$$

where $k = n\omega/c$ includes the index, n, which in turn includes the effect of the nonlinearity. The wave equation in this case takes the form

$$\frac{\partial^2 \mathscr{E}}{\partial z^2} - \mu_0\epsilon_0 n_0^2 \frac{\partial^2 \mathscr{E}}{\partial t^2} = \mu_0 \frac{\partial^2 \mathscr{P}_{NL}}{\partial t^2} \qquad (7.40)$$

where n_0 is the linear, or zero-field, refractive index. $\mathscr{E}$ as expressed in (7.24) and the nonlinear polarization in the form of (7.14) using (7.39) are substituted into the wave equation (using SVEA) to yield

$$-j2k\,\frac{dE_{01}}{dz} - k^2 E_{01} + \omega^2 \mu_0\epsilon_0 n_0^2 E_{01} = -\tfrac{3}{4}\,\omega^2\mu_0\epsilon_0\chi_{xxxx}|E_{01}|^2 E_{01} \qquad (7.41)$$

It is next assumed that the field amplitude is invariant with z, so that $dE_{01}/dz = 0$. Equation (7.41) is then solved for k^2, resulting in

$$k^2 = \omega^2 \mu_0 \epsilon_0 n^2 = \omega^2 \mu_0 \epsilon_0 \left(n_0^2 + \tfrac{3}{4} \chi_{xxxx} |E_{01}|^2 \right) \qquad (7.42)$$

The refractive index is then

$$n = n_0 \left(1 + \frac{3\chi_{xxxx}}{4n_0^2} |E_{01}|^2 \right)^{1/2} \approx n_0 + n_2' |E_{01}|^2 \qquad (7.43)$$

where a binomial expansion has been used to obtain the approximation, since $\chi_{xxxx} \ll 1$. The *nonlinear refractive index*, n_2', is identified as

$$n_2' = \frac{3\chi_{xxxx}}{8n_0} \qquad (7.44)$$

The change in index is thus seen to be dependent on the *intensity* of the optical field. The most important role played by the nonlinear index in fibers is in the phenomenon of *self-phase modulation* (SPM), in which the shape and frequency spectrum of a propagating pulse are influenced by the dynamic change in index that occurs in the pulse intensity envelope. These effects will be considered in the next section.

7.7. SELF-PHASE MODULATION

Consider an intense short light pulse that propagates in a third order material. Suppose that the nonlinearity is electronic; that is, its response time is extremely fast, enabling the medium (to a good approximation) to respond instantaneously to the intensity at each point in the pulse envelope. The medium refractive index will change with intensity as discussed in Section 7.6, but the change will be dynamic in the sense that each position along the pulse path undergoes a time-varying refractive index change that follows the pulse time envelope. Consider a plane wave pulse propagating in the z direction, having linearly polarized electric field:

$$\mathscr{E} = \tfrac{1}{2} E_0(z,t) \exp(-jnk_0 z) \exp(j\omega_0 t) + \text{c.c.} \qquad (7.45)$$

where $E_0(z,t)$ is the pulse envelope. It is important to note that a frequency measurement of this pulse *at any point in time* over the pulse envelope will always yield a band of frequencies centered at ω_0. At this stage, the intensity-dependent refractive index is included. This is expressed as

$$n = n_0 + n_2' |E_0(z,t)|^2 \qquad (7.46)$$

Substituting (7.46) into (7.45) yields

$$\mathscr{E} = \tfrac{1}{2} E_0(z,t) \exp(-jn_0 k_0 z) \exp[j\omega_0 t - j\delta\phi(t)] + \text{c.c.} \qquad (7.47)$$

where $\delta\phi(t) = n_2' k_0 |E_0(z,t)|^2 z$. The additional time-varying phase term, arising from the nonlinearity, has the effect of introducing new frequency components to the pulse that were not present before. The instantaneous frequency is found by differentiating the time-varying terms in the exponent:

$$\omega' = \frac{\partial}{\partial t}(\omega_0 t - \delta\phi(t)) = \omega_0 + \delta\omega(t) \qquad (7.48)$$

where

$$\delta\omega(t) = -\frac{\partial}{\partial t}\delta\phi(t) = -n_2' k_0 z \frac{\partial}{\partial t}|E_0(z,t)|^2 \qquad (7.49)$$

It is seen from the behavior of (7.49) that the added frequency content on the pulse is distributed over the pulse envelope in chirped fashion. The frequency offset as it relates to the pulse envelope is shown in Fig. 7.4.

A few points are to be noted with regard to the chirping behavior.

1. The frequency offset from the carrier is zero at the pulse center.
2. The maximum frequency offsets occur at the positions where the slope of the pulse is of maximum magnitude.
3. The increase in spectral width is in general inversely proportional to the pulse width for a given peak intensity, n_2', and medium length.
4. The spectrum of the self-phase modulated pulse is periodic, as illustrated in Fig. 7.5. The reason for the periodicity can be understood when noting that the chirped pulse will contain pairs of equal frequency components that are displaced from one another in time. A horizontal line can be drawn at any vertical position through the lower trace in Fig. 7.4 to demonstrate this. As a result, the phase separation between two equal frequency components will be frequency dependent. Constructive or destructive interference will thus occur as a function of frequency, leading to the periodic spectrum. The peaks on either end of the spectrum correspond to the inflection points on the pulse envelope.
5. If the medium response time is on the order of or slower than the pulse width, less spectral broadening will occur, since the material nonlinearity is unable to follow the pulse envelope in time, and thus produce sufficient phase modulation. For this reason, materials having electronic nonlinearities are the most favorable choice to maximize the effect.

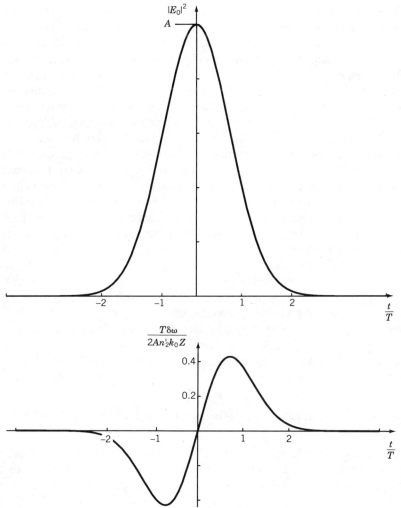

Figure 7.4. Plots of a gaussian pulse intensity envelope of the form $|E_0|^2 = A \exp(-t^2/T^2)$ (upper) and the frequency offset for the pulse as determined from (7.45) (lower) as functions of normalized time, t/T.

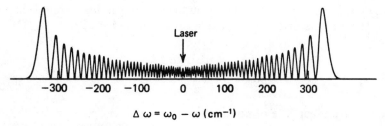

Figure 7.5. Frequency spectrum of a self-phase modulated pulse [3]. Copyright © 1984. Reprinted by permission of John Wiley & Sons, Inc.

Optical fibers introduce additional effects on a self-phase modulated pulse as a result of linear dispersion. As discussed in Chapter 5, the behavior of group index with wavelength in a single-mode fiber exhibits a typical U-shaped curve having its minimum at the zero-dispersion wavelength, λ_0. In the absence of self-phase modulation, pulses having frequency content corresponding to wavelengths less than λ_0 will be positively chirped as a result of normal linear dispersion. On the other hand, pulses that are tuned within the wavelength range to the right of λ_0 will be negatively chirped as a result of anomalous dispersion. In view of the positive chirping effect of self-phase modulation, it is seen that the combined effects of normal and nonlinear dispersion will result in substantially increased positive chirping of the pulse in addition to the frequency broadening; this will be the case at pulse wavelengths less than λ_0. At wavelengths greater than λ_0, however, the nonlinear and anomalous dispersions will cancel to an extent, leading to a pulse that will undergo less broadening as it propagates than would be true with either chirping effect acting alone. Propagation within either wavelength range produces results of considerable practical importance, as will be described in Chapter 8.

To fully describe the propagation of a pulse in a medium that exhibits both linear and nonlinear dispersion, these effects must be included in the wave equation. The development of this is demonstrated here in stages, in which the various effects are treated separately, and then later combined in a single equation [4].

To begin, consider a pulse that propagates in a medium in which there is zero dispersion and zero nonlinearity. The pulse electric field assumes the form

$$\mathscr{E}(z,t) = \tfrac{1}{2} E_0(z,t) \exp(j\omega_0 t - j\beta_0 z) + \text{c.c.} \qquad (7.50)$$

Since a plane wave is assumed, the propagation constant, β, is expressed in terms of the linear refractive index, n_0, which is assumed constant with frequency in the present case: $\beta_0 = n_0 k_0$. The wave equation will be

$$\frac{\partial^2 \mathscr{E}}{\partial z^2} - \mu_0 \epsilon_0 n_0^2 \frac{\partial^2 \mathscr{E}}{\partial t^2} = 0 \qquad (7.51)$$

Substituting (7.50) into (7.51) and using the slowly varying envelope approximation result in the first order equation for the pulse envelope:

$$\frac{\partial E_0(z,t)}{\partial z} + \frac{\beta_0}{\omega_0} \frac{\partial E_0(z,t)}{\partial t} = 0 \qquad (7.52)$$

where in the present case of no dispersion, $v_p = v_g = \omega_0/\beta_0$.

The inclusion of group velocity dispersion (GVD) will complicate matters since the velocity of each spectral component of the pulse will be a function of frequency. The pulse of (7.50) can be expressed in terms of the Fourier transform of the envelope through

$$\mathscr{E}(z,t) = \tfrac{1}{2} E_0(z,t) \exp(j\omega_0 t - j\beta_0 z) + \text{c.c.}$$

$$= \tfrac{1}{2} \int_{-\infty}^{\infty} A(\omega - \omega_0) \exp(j\omega t - j\beta z) d\omega + \text{c.c.} \qquad (7.53)$$

The pulse time envelope then becomes

$$E_0(z,t) = \int_{-\infty}^{\infty} A(\omega - \omega_0) \exp[j(\omega - \omega_0)t] \exp[-j(\beta - \beta_0)z] d(\omega - \omega_0)$$

$$(7.54)$$

The propagation constant is expanded, as before, to second order in a Taylor series about β_0:

$$\beta(\omega) = n_e(\omega)k_0 \approx \beta_0 + (\omega - \omega_0)\beta_1 + \tfrac{1}{2}(\omega - \omega_0)^2\beta_2 \qquad (7.55)$$

Substituting (7.55) into (7.54) results in

$$E_0(z,t) = \int_{-\infty}^{\infty} A(\omega - \omega_0) \exp[j(\omega - \omega_0)t]$$

$$\cdot \exp[-j(\omega - \omega_0)\beta_1 z] \exp[-j\tfrac{1}{2}(\omega - \omega_0)^2\beta_2 z] d(\omega - \omega_0)$$

$$(7.56)$$

The procedure from here would ordinarily be to substitute (7.56) into the wave equation given by (7.51) and to apply the slowly varying envelope approximation. The result would be a first order differential equation for $E_0(z,t)$. Instead, it is possible simply to differentiate (7.56) with respect to z and t, and to identify terms. The result is

$$\frac{\partial E_0(z,t)}{\partial z} + \beta_1 \frac{\partial E_0(z,t)}{\partial t} = j \frac{\beta_2}{2} \frac{\partial^2 E_0(z,t)}{\partial t^2} \qquad (7.57)$$

The left side of the above is seen to be the same as that of (7.52), except for the β_1 term, identified as the reciprocal of the group velocity at ω_0.

Finally, consider a nonlinear medium with no linear dispersion. The effective index is expressed in terms of the pulse field using (7.46). Then, assuming n'_2 is small, it is found that $n^2 \approx n_0^2 + 2n_0 n'_2 |E_0(z,t)|^2$. On substitution, the wave equation becomes

$$\frac{\partial^2 \mathscr{E}}{\partial z^2} - \mu_0 \epsilon_0 n_0^2 \frac{\partial^2 \mathscr{E}}{\partial t^2} = 2\mu_0 \epsilon_0 n_0 n_2' \frac{\partial^2}{\partial t^2}(|E_0|^2 \mathscr{E}) \qquad (7.58)$$

Substituting (7.50) and invoking the slowly varying envelope approximation result in the following first order equation for the pulse envelope:

$$\frac{\partial E_0(z,t)}{\partial z} + \beta_1 \frac{\partial E_0(z,t)}{\partial t} = -j \frac{n_2'}{n_0} \beta_0 |E_0(z,t)|^2 E_0(z,t) \qquad (7.59)$$

where in the case of no linear dispersion $\beta_1 = n_0 \sqrt{\mu_0 \epsilon_0} = \beta_0/\omega_0$.

Since, in general, linear and nonlinear dispersions are relatively weak effects (β_2 and n_2' are small), the combined effects of the two can be accounted for by a simple addition of the right-hand sides of (7.57) and (7.59). Losses can also be included by adding a third term to the right-hand side proportional to the power loss coefficient, α. Saturable gain (as would be the case in a fiber amplifier) is included in a similar fashion. The result of the above reasoning is the *nonlinear Schrödinger equation*, describing nonlinear pulse propagation in a dispersive medium:

$$\frac{\partial E_0(z,t)}{\partial z} + \beta_1 \frac{\partial E_0(z,t)}{\partial t} = -\frac{\alpha}{2} E_0(z,t)$$
$$+ j \frac{\beta_2}{2} \frac{\partial^2 E_0(z,t)}{\partial t^2} - j \frac{n_2'}{n_0} \beta_0 |E_0(z,t)|^2 E_0(z,t) \qquad (7.60)$$

It is convenient to transform (7.60) into a moving reference frame that follows the pulse peak as it propagates at group velocity v_{g0} (Problem 7.6). To accomplish this, a local time parameter, $t_L = t - z/v_{g0} = t - \beta_1 z$, is defined. The transformed nonlinear Schrödinger equation, with contributions from the various effects identified, thus reads

$$\frac{\partial E_0(z,t_L)}{\partial z} = \underbrace{-\frac{\alpha}{2} E_0(z,t_L)}_{\text{Loss}} + \underbrace{j \frac{\beta_2}{2} \frac{\partial^2 E_0(z,t_L)}{\partial t_L^2}}_{\text{GVD}} - \underbrace{j \frac{n_2'}{n_0} \beta_0 |E_0(z,t_L)|^2 E_0(z,t_L)}_{\text{SPM}}$$

$$(7.61)$$

In the case of lossless propagation ($\alpha = 0$), solutions of (7.61) exist that are invariant over time and propagation distance. These solutions are known as *optical solitons*. The physical significance of a soliton is that its shape, amplitude, and center frequency are such that the induced chirp on the pulse that arises from self-phase modulation is exactly canceled by the chirping due to group velocity dispersion. The pulse shape that is necessary to achieve this is the hyperbolic secant. The time-invariant solution to (7.61), shown below, is readily verified by substitution (Problem 7.8):

$$E_0(z, t_L) = A_0 \exp\left(j \frac{\beta_2 z}{2T^2}\right) \text{sech}\left(\frac{t_L}{T}\right) \qquad (7.62)$$

where

$$|A_0|^2 = \frac{-n_0\beta_2}{n_2'\beta_0 T^2} \qquad (7.63)$$

It is seen that for (7.62) to be a solution, the product of the pulse amplitude A_0 and pulse width parameter T will be a constant, as observed in (7.63). A shorter pulse requires a greater amplitude to propagate as a soliton.

It is apparent from (7.63) that for a soliton to exist, n_2' and β_2 must be of *opposite sign*; in other words, in silica fibers (where n_2' is positive), the spectral content of the pulse must lie within the wavelength range *greater* than the zero dispersion wavelength, λ_0. This enables the negative-chirping action of the fiber to counteract the positive-chirping effect of self-phase modulation. The specific application of (7.61) to silica fibers will be addressed in Chapter 8.

7.8. STIMULATED RAMAN SCATTERING

At power levels on the order of 100 mW in long fibers, the generation of additional forward-propagating light at down-shifted frequencies may occur through stimulated Raman scattering (SRS). The Raman effect arises from vibrational oscillations that occur between constituent atoms within individual molecules or structures of molecules. The resonant frequencies are considerably lower than optical frequencies, due to the relatively large masses of the atoms involved. This is to be distinguished from electronic oscillations, in which the electron position over time occurs in step with the applied optical field orientation.

Stimulated Raman scattering occurs when a slow resonance is excited by two optical fields at two frequencies that differ by an amount that is in the vicinity of a molecular resonant frequency. The optical fields coherently interact to produce sum and difference frequencies. The interaction occurs by way of the product of the fields—thus making the process nonlinear. The result is a driving force at the difference frequency that excites the resonances. The interaction between the fields and the oscillators results in a transfer of power between the two optical waves. In principle, either down-shifted (Stokes) or up-shifted (anti-Stokes) light may be generated through the interaction, but in the case of fiber waveguides in which all waves are constrained to propagate in the same direction, the Stokes process dominates. This is because Stokes wave generation (1) does not require previous excitation of the resonances and (2) is phase matched for collinear propagation of the two waves. Aside from being phase mismatched for collinear propagation, the anti-Stokes process involves the interaction of the input light with Raman oscillators that have been *previously excited*. Since the population of excited molecules is usually much less than that of the unexcited ones, the anti-Stokes radiation is, for that reason alone, generally much weaker than Stokes light.

Stokes wave generation can begin with a single input field, referred to as the pump wave. The down-shifted light originates by way of *spontaneous* Raman scattering, which occurs as a result of a slight excitation of the molecular resonances. As the scattered Stokes light increases in intensity, the stimulated scattering regime is eventually reached; in this case, the Stokes light interacts with the pump light to further excite the resonances, and thus the rate of frequency conversion can be dramatically increased.

A classical model of the Raman effect accounts for the nonlinearity through a term describing the change in polarizability with atomic displacement from equilibrium within a molecule. Consider the polarization

$$\mathscr{P} = \epsilon_0 \chi \mathscr{E} = \epsilon_0 N \alpha_p \mathscr{E} \tag{7.64}$$

where N is the number density of molecules and α_p is the molecular polarizability (associated with a single molecule). The nonlinear part of the polarization is obtained by expanding α_p into first and higher order terms. This is accomplished by recognizing that α_p will depend on the relative positions of the atoms within the molecule at a given time. If the atoms are displaced to positions away from equilibrium, then the molecule will polarize to a different extent, when an external field is applied, than it does when the component atoms are at their equilibrium positions.

Consider the case in which a vibrational oscillation mode is such that atoms are periodically displaced from equilibrium along a single direction. The displacement of a single atom, $q(z,t)$, (treated as a scalar in this analysis) could be represented by the function $q(z,t) = Q(z)\cos(\omega_{0r}t)$, where ω_{0r} is the resonant (or Raman) frequency. z is the propagation direction of both optical fields. The oscillation mode is termed *Raman active* if the atomic displacement results in no change in the dipole moment of the molecule. An example of a Raman active mode is the symmetric oscillation of two identical atoms on opposite sides of a central atom, in which the oscillation occurs along the line joining the three, and where the two outer atoms are always equidistant from the center. The resulting time-varying polarizability can be described by the expansion

$$\alpha_p(z,t) = \alpha_{p0} + q(z,t)\left(\frac{\partial \alpha_p}{\partial q}\right)_{q=0} = \alpha_{p0} + \alpha_{p1}q(z,t) \tag{7.65}$$

where $\alpha_{p1} = \partial \alpha_p / \partial q|_{q=0}$ is the change in polarizability with displacement, evaluated at the equilibrium position. When (7.65) is substituted into (7.64), it is seen that the time-varying displacement will give rise to polarization terms that are up- and down-shifted in frequency from that of the input light by the Raman frequency.

The nonlinear part of the polarization expressed in (7.64), using (7.65), is found to be

$$\mathscr{P}_{NL} = \epsilon_0 N \alpha_{p1} q(z,t) \mathscr{E}(z,t) \tag{7.66}$$

$\mathscr{E} = \mathscr{E}_1 + \mathscr{E}_2$ is the total field, expressed as the sum of the Stokes and pump fields, respectively. The latter fields are given by

$$\mathscr{E}_i(z,t) = \tfrac{1}{2}\mathbf{E}_{0i}\exp(-jk_iz)\exp(j\omega_it) + \text{c.c.} = \tfrac{1}{2}\mathbf{E}_i\exp(j\omega_it) + \text{c.c.} \tag{7.67}$$

where $k_i = k^{\omega i}$ ($i = 1, 2$), and where the pump frequency, ω_2, is greater than the Stokes frequency, ω_1.

The displacement, $q(z,t)$, is expected to be proportional to the difference frequency term in the scalar product of $\mathscr{E}_1$ and $\mathscr{E}_2$. The actual form of $q(z,t)$ can be determined by solving an equation of motion for the atom involved, based on (1.52) [1]:

$$m\frac{\partial^2 q}{\partial t^2} + m\zeta_r\frac{\partial q}{\partial t} + k_s q = F(z,t) \tag{7.68}$$

where m is the mass of the atom, ζ_r is a phenomenological damping coefficient, and k_s is the "spring constant" describing the restoring force on the atom and giving rise to the resonant frequency $\omega_{0r} = \sqrt{k_s/m}$. The forcing function, $F(z,t) = -e\mathscr{E}$, was used in driving the electronic oscillations in (1.52). In the present case, a forcing function in the form of the above field product is anticipated. The function can be determined by using the fact that force is the gradient of the stored energy [1]. The latter is found through the energy density in the electric field:

$$w_e = \tfrac{1}{2}\epsilon\mathscr{E}\cdot\mathscr{E} \quad \text{J/m}^3 \tag{7.69}$$

where

$$\epsilon = 1 + \chi = \epsilon_0\{1 + N[\alpha_{p0} + \alpha_{p1}q(z,t)]\} \tag{7.70}$$

The gradient of w_e is thus the force per unit volume, meaning that the force per oscillator is

$$F(z,t) = \frac{1}{N} \frac{\partial w_e}{\partial q} \tag{7.71}$$

The energy density is evaluated by substituting the total field, $\mathscr{E} = \mathscr{E}_1 + \mathscr{E}_2$, into (7.69). In evaluating $\mathscr{E} \cdot \mathscr{E}$, only the difference frequency terms (involving $\omega_2 - \omega_1$) are retained. This is because the molecular resonator can respond only to low frequencies, where ω_1 is close to ω_2. Additionally, since the scalar product of the fields is involved, only field components having the same polarization can take part in the interaction. Using (7.67) and (7.70) in (7.69) with the above restrictions, (7.71) finally becomes

$$F(z,t) = \frac{\epsilon_0 \alpha_{p1}}{8} \{ E_{02} E_{01}^* \exp[-j(k_2 - k_1)z] \exp[j(\omega_2 - \omega_1)t] + c.c. \} \tag{7.72}$$

The above forcing function is now substituted into the equation of motion, (7.68), along with the assumed product solution form for the displacement function:

$$q(z,t) = \tfrac{1}{2} Q(z) \exp[j(\omega_2 - \omega_1)t] + c.c. \tag{7.73}$$

The result is

$$q(z,t) = \frac{\epsilon_0 \alpha_{p1}}{8m} \frac{E_{02} E_{01}^* \exp[-j(k_2 - k_1)z] \exp[j(\omega_2 - \omega_1)t]}{[\omega_{0r}^2 - (\omega_2 - \omega_1)^2 + j\zeta_r(\omega_2 - \omega_1)]} + c.c. \tag{7.74}$$

The nonlinear polarization is now constructed by substituting (7.74) into (7.66). Of interest are the polarizations that oscillate at the pump and Stokes frequencies. The polarization at ω_1 is found by keeping two of the terms in the product $q(z,t)\mathscr{E}_2$ in (7.66). The result is

$$\mathscr{P}_{NL}^{\omega_1}(z,t) = \frac{\epsilon_0^2 N \alpha_{p1}^2}{16m} \left(\frac{|E_{02}|^2 E_{01} \exp(-jk_1 z) \exp(j\omega_1 t)}{[\omega_{0r}^2 - (\omega_2 - \omega_1)^2 - j\zeta_r(\omega_2 - \omega_1)]} + c.c. \right) \tag{7.75}$$

The polarization at ω_2 is obtained from two terms in the product $q(z,t)\mathscr{E}_1$ in (7.66):

$$\mathscr{P}_{NL}^{\omega_2}(z,t) = \frac{\epsilon_0^2 N \alpha_{p1}^2}{16m} \left(\frac{|E_{01}|^2 E_{02} \exp(-jk_2 z) \exp(j\omega_2 t)}{[\omega_{0r}^2 - (\omega_2 - \omega_1)^2 + j\zeta_r(\omega_2 - \omega_1)]} + c.c. \right) \tag{7.76}$$

The other terms in (7.66), yielding polarizations that oscillate at frequencies other than ω_1 and ω_2, describe interactions that are phase mismatched for collinear propagation; these will give rise to extremely weak fields and will have negligible effect on the evolution of the pump and Stokes waves. All terms in (7.66) involve the product of three fields, thus identifying all Raman processes as third order. Note that the propagation constants for the polarizations in (7.75) and (7.76) are those associated with the polarization frequencies, that is, k_1 and k_2. Both polarizations thus describe processes that are phase matched.

The polarization of (7.75) is identified with Stokes wave generation, since it oscillates at frequency ω_1. Equation (7.76) becomes important as the Stokes field amplitude E_{01} becomes appreciable. This process represents the back-transfer of power from the Stokes wave to the pump and is referred to as the *inverse Raman effect* [5]. Both polarizations must be included in the wave equation to account for growth and depletion of the pump and Stokes fields.

Equations (7.75) and (7.76) can be written in the form

$$\mathscr{P}_{NL}^{\omega_i}(z,t) = \tfrac{1}{2} P_{NL}^{\omega_i} \exp(j\omega_i t) + \text{c.c.} \tag{7.77}$$

where $i = 1$ or 2, and where the polarization phasors are

$$P_{NL}^{\omega_1} = \epsilon_0 \chi_r^{\omega_1} |E_{02}|^2 E_{01} \exp(-jk_1 z) \tag{7.78a}$$
$$P_{NL}^{\omega_2} = \epsilon_0 \chi_r^{\omega_2} |E_{01}|^2 E_{02} \exp(-jk_2 z) \tag{7.78b}$$

Using the above with (7.75) and (7.76), the Raman susceptibilities are identified. The susceptibility for the Stokes wave is:

$$
\begin{aligned}
\chi_r^{\omega_1} &= \frac{\epsilon_0 N \alpha_{p1}^2}{8m} \left(\frac{[\omega_{0r}^2 - (\omega_2 - \omega_1)^2] + j\zeta_r(\omega_2 - \omega_1)}{[\omega_{0r}^2 - (\omega_2 - \omega_1)^2]^2 + \zeta_r^2(\omega_2 - \omega_1)^2} \right) \\
&\approx -\frac{\epsilon_0 N \alpha_{p1}^2}{8m\omega_{0r}\zeta_r} \left(\frac{-j + \delta_r}{1 + \delta_r^2} \right)
\end{aligned}
\tag{7.79}
$$

where the normalized detuning parameter is $\delta_r \equiv (2/\zeta_r)[(\omega_2 - \omega_1) - \omega_{0r}]$. The approximation in (7.79) is made under the assumption of near-resonance operation, such that $\omega_2 - \omega_1 \approx \omega_{0r}$, as was used to approximate (1.56) by (1.57). This approximation is valid over several normalized linewidths, δ_r, away from resonance, provided $\zeta_r \ll \omega_{0r}$. The susceptibility associated with the pump wave is found to be the complex conjugate of (7.79):

$$\chi_r^{\omega_2} = \chi_r^{\omega_1*} \approx -\frac{\epsilon_0 N \alpha_{p1}^2}{8m\omega_{0r}\zeta_r} \left(\frac{j + \delta_r}{1 + \delta_r^2} \right) \tag{7.80}$$

Plots of the real and imaginary parts of either of the above susceptibilities will thus be of the same form as those shown in Fig. 1.5, except the horizontal axis in that figure will indicate the frequency difference, $\omega_2 - \omega_1$. An important consequence of the fact that (7.79) and (7.80) are complex conjugates of each other is that the imaginary part of (7.80) is associated with attenuation for the pump wave (consistent with the behavior of (1.61)), whereas the imaginary part of (7.79), being of opposite sign, is associated with *gain* for the Stokes wave in the presence of the pump.

The fields and polarizations are now substituted into the wave equation (7.6), using $\mathscr{P}_{NL} = \mathscr{P}_{NL}^{\omega 1} + \mathscr{P}_{NL}^{\omega 2}$, $\mathscr{E} = \mathscr{E}_1 + \mathscr{E}_2$, and where all fields and polarizations are linearly polarized in the same direction. Proceeding in a manner similar to that described in Section 7.6, including the use of SVEA, the wave equation separates into two coupled equations that describe the evolution with distance of the Stokes and pump field amplitudes:

$$\frac{dE_{01}}{dz} = -\frac{j\omega_1}{2nc}\,\chi_r|E_{02}|^2 E_{01} \qquad (7.81a)$$

$$\frac{dE_{02}}{dz} = -\frac{j\omega_2}{2nc}\,\chi_r^*|E_{01}|^2 E_{02} \qquad (7.81b)$$

where n is the refractive index (assumed to be the same at both frequencies) and where $\chi_r = \chi_r^{\omega 1}$.

The coupled equations can be rewritten in terms of the wave intensities by performing the following operations, shown below for (7.81a):

$$E_{01}^* \frac{dE_{01}}{dz} + E_{01}\frac{dE_{01}^*}{dz} = \frac{d|E_{01}|^2}{dz} = (-j\chi_r + j\chi_r^*)\,\frac{\omega_1}{2nc}|E_{02}|^2|E_{01}|^2 \quad (7.82)$$

The wave intensity (in W/m^2) is found for either frequency through $I_i = (n/2\eta_0)|E_{0i}|^2$. Using this in (7.82) and then performing a similar procedure for (7.81b), we find

$$\frac{dI_1}{dz} = g_r I_1 I_2 - \alpha I_1 \qquad (7.83a)$$

$$\frac{dI_2}{dz} = -\frac{\omega_2}{\omega_1}g_r I_1 I_2 - \alpha I_2 \qquad (7.83b)$$

In obtaining these, the terms αI_1 and αI_2 have been added phenomenologically to account for linear losses in the medium; the losses are characterized by exponential power attenuation coefficient α, which is assumed the same at both wavelengths.

The *Raman gain coefficient*, g_r, is

$$g_r = \frac{\epsilon_0 N \eta_0 \omega_1 \alpha_{p1}^2}{4m\omega_{0r}\zeta_r cn^2} \left(\frac{1}{1+\delta_r^2} \right) \tag{7.84}$$

g_r is typically determined through measurement. Equation (7.84) is useful in that it shows the general form of g_r and its behavior with the various parameters—most notably its linear increase with the optical frequency ω_1, and its proportionality to the lineshape function, $1/(1+\delta_r^2)$. In practice, more complicated lineshapes are usually encountered; these are accounted for by replacing $1/(1+\delta_r^2)$ in (7.84) with the actual lineshape function.

The solutions of (7.83a) and (7.83b) are (as can be verified by direct substitution)

$$I_1(z) = \frac{\omega_1}{\omega_2} I_0 \exp(-\alpha z) \frac{\psi_r}{1+\psi_r} \tag{7.85}$$

$$I_2(z) = \frac{I_0 \exp(-\alpha z)}{1+\psi_r} \tag{7.86}$$

where

$$\psi_r = \frac{\omega_2}{\omega_1} \frac{I_{10}}{I_{20}} \exp\left(\frac{g_r I_0 [1 - \exp(-\alpha z)]}{\alpha} \right) \tag{7.87}$$

I_{10} and I_{20} are the input ($z = 0$) intensities of the Stokes and pump waves, and $I_0 = I_{20} + (\omega_1/\omega_2)I_{10}$.

One special case is that in which the input Stokes wave is weak, such that $I_{10} \ll I_{20}$, and so $\psi_r \ll 1$. Equations (7.85) and (7.86) become

$$I_1(z) \approx I_{10} \exp(-\alpha z) \exp\left(\frac{g_r I_{20}[1 - \exp(-\alpha z)]}{\alpha} \right) \tag{7.88}$$

and

$$I_2(z) \approx I_{20} \exp(-\alpha z) \tag{7.89}$$

These are valid as long as the Stokes wave, during amplification, remains weak compared to the pump. These results do not include the effects of spontaneous Raman scattering, which are important when $I_{10} = 0$. In this case, an effective Stokes input intensity is used in place of I_{10} in (7.88), as will be described in Chapter 8. There, the present theory is applied to the specific problem of Raman scattering in fibers.

7.9. STIMULATED BRILLOUIN SCATTERING

Stimulated Brillouin scattering (SBS) is a process that is analogous to stimulated Raman scattering, in that the end result is a wave (again referred to as the Stokes signal) that is down-shifted in frequency from that of a strong pump wave. The interaction between the two fields occurs, however, by way of an acoustic wave that propagates codirectionally with the pump, instead of through a slow material resonance, as was the case with Raman scattering. The acoustic wave is initiated by the pump electric field, which causes changes in the material density though *electrostriction*. A material density wave results that propagates at the velocity of sound in the medium. If propagation is constrained along a single dimension (as is the case in a fiber), a pressure wave (sound wave) is established that moves in the direction of the pump. Since refractive index changes with material density, the periodic changes in density associated with the pressure wave are manifest as a moving refractive index grating. This structure diffracts a portion of the pump wave in the backward direction, resulting in the Stokes signal. The down-shift in signal frequency arises from the Doppler shift that occurs as the pump diffracts from the moving grating.

The process becomes stimulated as the Stokes wave increases in amplitude; it then participates in the nonlinear process by interacting with the pump to reinforce the acoustic wave. This can be visualized by considering the interference pattern that results from two counterpropagating fields having different frequencies, as was studied in Problem 1.10. The intensity fringes associated with the interference pattern will move at a velocity proportional to the frequency difference. If the frequency difference is adjusted such that the fringe velocity matches that of sound in the material, then an acoustic wave of wavelength equal to the fringe spacing can be initiated and reinforced by the two optical fields. If this situation exists, then the coupling between the optical waves is said to be *Brillouin enhanced*. Interestingly, the Stokes wavefront is proportional to the *phase conjugate* of that of the pump, leading to applications involving the correction of aberrations in optical systems. Discussions of this and related topics are found in [1].

The wave geometry for backward stimulated Brillouin scattering is shown in Fig. 7.6. The pump, Stokes, and sound frequencies are ω_2, ω_1, and ω_p, respectively; the waves are further described by propagation vectors, $\mathbf{k}_2 = k_2\hat{\mathbf{a}}_z$, $\mathbf{k}_1 = -k_1\hat{\mathbf{a}}_z$, and $\mathbf{k}_p = k_p\hat{\mathbf{a}}_z$. The fields (assumed co-polarized) and the acoustic wave take the forms

$$\mathscr{E}_1(z,t) = \tfrac{1}{2} E_{01}(z) \exp[j(\omega_1 t + k_1 z)] + \text{c.c.} \tag{7.90}$$

$$\mathscr{E}_2(z,t) = \tfrac{1}{2} E_{02}(z) \exp[j(\omega_2 t - k_2 z)] + \text{c.c.} \tag{7.91}$$

$$q(z,t) = \tfrac{1}{2} Q(z) \exp[j(\omega_p t - k_p z)] + \text{c.c.} \tag{7.92}$$

where $Q(z)$ is the displacement from equilibrium of a small volume in the

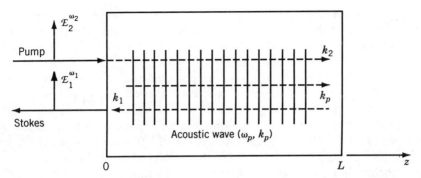

Figure 7.6. Beam and wave geometry for backward stimulated Brillouin scattering.

medium (considered a point) as a result of the applied electrostrictive force (this is to be distinguished from the interpretation of Q in Raman scattering as the displacement of a specific atom or molecule). Such displacement results in a local change in material density. Equation (7.88) thus describes a propagating density wave, although it is expressed in terms of displacement.

The complex nonlinear polarization that radiates the Stokes wave is anticipated to be of the form

$$P_{NL}^{\omega_1} \propto Q^* E_{02} \exp\left[-j(\omega_p t - k_p z)\right] \exp\left[j(\omega_2 t - k_2 z)\right]$$
$$\propto E_{01} E_{02}^* E_{02} \exp\left[j(\omega_1 t + k_1 z)\right] \tag{7.93}$$

where $\omega_p = \omega_2 - \omega_1$, and where, under phase-matched conditions, the **k**-vector magnitudes are related through $k_p = k_1 + k_2$. To determine ω_p and ω_1, knowing ω_2, the assumption that $\omega_p \ll \omega_1, \omega_2$ is used. It follows that $k_p \approx 2k_2$. Then, using $k_p = \omega_p/v_p$, where v_p is the sound velocity, the sound frequency is $\omega_p \approx 2n^{\omega_2}\omega_2 v_p/c$. Consequently, the Stokes frequency under phase-matched operation is

$$\omega_1 = \omega_{10} = \omega_2\left(1 - 2n\frac{v_p}{c}\right) \tag{7.94}$$

The displacement function is governed by an equation of motion that is in direct analogy to (7.68) [1]:

$$\rho\frac{\partial^2 q}{\partial t^2} + \rho\varsigma\frac{\partial q}{\partial t} - T_e\frac{\partial^2 q}{\partial z^2} = \frac{\gamma}{2}\frac{\partial \mathscr{E}_T^2}{\partial z} \tag{7.95}$$

where ρ is the mass density, ς is a phenomenological damping coefficient, and $\mathscr{E}_T = \mathscr{E}_1 + \mathscr{E}_2$. The third term in (7.95) represents the force per unit volume acting to restore the medium to zero displacement. The force is determined through T_e, the bulk modulus of elasticity, which relates the change in material density to the applied pressure. Finally, γ is the electrostrictive coefficient, measuring the change in medium permittivity with density through $\gamma = \rho(\partial\epsilon/\partial\rho)$. The forcing function appearing on the right-hand side of (7.95) is derived by relating the force per unit volume in the material to the change in stored energy in the electric fields. The arguments are analogous to those used in obtaining (7.71) for stimulated Raman scattering and are explained in detail in [1]. For the present study, the important result is that the force (and hence the degree of medium displacement) is related to the product of two fields, thus justifying the relations in (7.93) and establishing SBS as a third order nonlinear process.

Rearranging terms, (7.95) assumes the more familiar form of a wave equation:

$$\frac{\partial^2 q}{\partial z^2} = \frac{1}{v_p^2}\frac{\partial^2 q}{\partial t^2} + \frac{\zeta}{v_p^2}\frac{\partial q}{\partial t} - \frac{1}{2}\frac{\gamma}{\rho v_p^2}\frac{\partial \mathscr{E}_T^2}{\partial z} \tag{7.96}$$

where the sound velocity, $v_p = \sqrt{T_e/\rho}$. Using (7.92) and invoking the slowly-varying envelope approximation for Q, the left-hand side of (7.96) becomes:

$$\frac{\partial^2 q}{\partial z^2} \approx -\frac{1}{2}\left[k_p^2 Q + j2k_p\frac{dQ}{dz}\right]\exp[j(\omega_p t - k_p z)] + \text{c.c.} \tag{7.97}$$

When $\mathscr{E}_T^2$ is evaluated, only terms that oscillate at $\omega_p = \omega_2 - \omega_1$ are retained. With these conditions, (7.90)–(7.92) and (7.97) are substituted into (7.96) to result in:

$$-\frac{1}{2}\left(k_p^2 Q + j2k_p\frac{dQ}{dz}\right)\exp[j(\omega_p t - k_p z)]$$

$$= \left(-\frac{\omega_p^2}{2v_p^2} + j\frac{\omega_p \zeta}{2v_p^2}\right)Q\exp[j(\omega_p t - k_p z)]$$

$$+ j\frac{\gamma k_p}{4\rho v_p^2}E_{02}E_{01}^*\exp[j(\omega_p t - (k_2 + k_1)z)] \tag{7.98}$$

where the approximation $|\partial/\partial z(E_{01}^* E_{02})| \ll |(k_1 + k_2)E_{01}^* E_{02}|$ has been used. Noting that $k_p = \omega_p/v_p$, (7.98) simplifies to

$$\frac{dQ}{dz} = -\frac{\gamma}{4\rho v_p^2}E_{02}E_{01}^*\exp(-j\Delta k z) - \frac{\alpha_p}{2}Q \tag{7.99}$$

where the attenuation coefficient for the pressure wave is $\alpha_p = \zeta/v_p$.

The phase mismatch per unit distance appearing in (7.99) is defined as $\Delta k \equiv k_1 + k_2 - k_p$. Deviations of the Stokes frequency from ω_{10}, specified by (7.94), will cause an increase in the magnitude of Δk. Writing this as $\Delta k = (n\omega_1/c) - (\omega_p/v_p) + (n\omega_2/c)$, and using $\omega_2 - \omega_1 = \omega_p$, along with the approximation $\omega_1 \approx \omega_2$, the phase mismatch becomes [6]

$$\Delta k \approx \frac{1}{v_p}(\omega_1 - \omega_{10}) \tag{7.100}$$

The ultimate effect of Δk is a decrease in the Brillouin gain as ω_1 is detuned from ω_{10}, as will be shown.

The nonlinear polarizations that radiate the optical fields are found through the change in medium permittivity resulting from the strain produced by the pressure wave. For a strain in one dimension (dq/dz), this change will be

$$\Delta\epsilon = -\gamma \frac{\partial q}{\partial z} \tag{7.101}$$

To obtain the nonlinear polarization, the relation between polarization, electric displacement, and permittivity can be used:

$$\mathcal{D} = \epsilon\mathcal{E} = \epsilon_0\mathcal{E} + \mathcal{P}_L + \mathcal{P}_{NL} \tag{7.102}$$

Using (7.101), along with the relation $\epsilon = \epsilon_L + \Delta\epsilon$, the nonlinear polarization is in general given by [1]

$$\mathcal{P}_{NL} = -\gamma \frac{\partial q}{\partial z} \mathcal{E}_T \tag{7.103}$$

Equations (7.90) to (7.92) and (7.103) are next substituted into the wave equation for z propagation, (7.29), which in turn separates into an equation at ω_1 and another at ω_2. Using the slowly varying envelope equation, the ω_1 equation will be

$$jk_1 \frac{dE_{01}}{dz} \exp\left[j(\omega_1 t + k_1 z)\right] + \text{c.c.} = \mu_0 \frac{\partial^2 \mathcal{P}_{NL}^{\omega_1}}{\partial t^2} \tag{7.104}$$

The nonlinear polarization in this equation is the term in (7.103) that oscillates at ω_1:

$$\mathcal{P}_{NL}^{\omega_1} = -j \frac{\gamma k_p}{4} E_{02} Q^* \exp(j\omega_1 t) \exp\left[-j(k_2 - k_p)z\right] + \text{c.c.} \tag{7.105}$$

where $|dQ^*/dz|$ is assumed much less than $|k_p Q^*|$. Substituting (7.105) into (7.104) results in the following equation describing the growth of the Stokes field:

$$\frac{dE_{01}}{dz} = \frac{\gamma k_p k_1}{4\epsilon_0 n^2} E_{02} Q^* \exp(-j\Delta kz) + \frac{\alpha}{2} E_{01} \qquad (7.106)$$

where $k_1 = n\omega_1 \sqrt{\mu_0 \epsilon_0}$. Following a similar procedure, the wave equation at ω_2 reduces to

$$\frac{dE_{02}}{dz} = \frac{\gamma k_p k_2}{4\epsilon_0 n^2} E_{01} Q \exp(j\Delta kz) - \frac{\alpha}{2} E_{02} \qquad (7.107)$$

In (7.106) and (7.107), the possibility of exponential loss has been included by adding the terms involving a power loss coefficient, α, which is assumed to be the same at both frequencies.

Equations (7.99), (7.106), and (7.107) form a set of coupled equations that describe the spatial evolution of pump, signal, and acoustic waves. It is possible to convert these three to a set of two equations for the fields, analogous to (7.81a) and (7.81b) for SRS, by assuming that the acoustic wave is heavily damped [6]. The procedure is to first multiply (7.99) by $\exp(\alpha_p z/2)$ and then integrate both sides of the equation from 0 to z:

$$\int_0^z \frac{d}{dz'}\left[Q \exp\left(\frac{\alpha_p z'}{2}\right)\right] dz'$$

$$= -\frac{\gamma}{4\rho v_p^2} \int_0^z E_{02}(z')E_{01}^*(z') \exp(-j\Delta kz') \exp\left(\frac{\alpha_p z'}{2}\right) dz' \quad (7.108)$$

where the relation $(dQ/dz)\exp(\alpha_p z/2) + (\alpha_p/2)Q\exp(\alpha_p z/2) = d/dz(Q\exp(\alpha_p z/2))$ has been used. Carrying out the integral on the left-hand side and solving for $Q(z)$, we obtain

$$Q(z) = -\frac{\gamma}{4\rho v_p} \int_0^z E_{02}(z')E_{01}^*(z') \exp(-j\Delta kz')$$

$$\cdot \exp\left(-\frac{\alpha_p}{2}(z-z')\right) dz' + Q(0) \exp\left(-\frac{\alpha_p z}{2}\right) \qquad (7.109)$$

As described in [6], the procedure for evaluating this integral is to note that the primary contribution will be over the range $(z-z') \le 2/\alpha_p$. It is further assumed that the acoustic wave damping rate is much higher than the rates at which either of the two fields change. Consequently, E_{01}^* and E_{02} are approximated as constant over the range of integration and are evaluated at $z' = z$. The result is

$$Q(z) = -\frac{\gamma}{4\rho v_p} \left(\frac{\exp(-j\Delta kz) - \exp(-\alpha_p z/2)}{(\alpha_p/2 - j\Delta k)} \right) E_{02}(z)E_{01}^*(z)$$
$$+ Q(0) \exp(-\alpha_p z/2) \tag{7.110}$$

Next, (7.110) is substituted into (7.106) and (7.107) under the assumptions that $Q(0) = 0$ and $\exp(-\alpha_p z/2) \ll 1$. The results are

$$\frac{dE_{01}}{dz} = -\frac{\gamma^2 k_p k_1}{16\epsilon_0 n^2 \rho v_p^2} \frac{(\alpha_p/2 - j\Delta k)}{[\alpha_p^2/4 + (\Delta k)^2]} |E_{02}(z)|^2 E_{01}(z)$$
$$+ \frac{\alpha}{2} E_{01}(z) \tag{7.111}$$

$$\frac{dE_{02}}{dz} = -\frac{\gamma^2 k_p k_2}{16\epsilon_0 n^2 \rho v_p^2} \frac{(\alpha_p/2 + j\Delta k)}{[\alpha_p^2/4 + (\Delta k)^2]} |E_{01}(z)|^2 E_{02}(z)$$
$$- \frac{\alpha}{2} E_{02}(z) \tag{7.112}$$

The above equations are expressed in terms of the wave intensities by performing the same procedure that was demonstrated in (7.82) for stimulated Raman scattering. Using $|E_{0i}|^2 = (2\eta_0/n)I_i$, where $i = 1, 2$, the intensity equations assume forms that are analogous to (7.83a) and (7.83b):

$$\frac{dI_1}{dz} = -g_b I_1 I_2 + \alpha I_1 \tag{7.113a}$$

$$\frac{dI_2}{dz} = -g_b I_1 I_2 - \alpha I_2 \tag{7.113b}$$

The *Brillouin gain*, g_b, is defined as

$$g_b = \frac{\gamma_2 k_p k_2 \eta_0 \alpha_p}{2\epsilon_0 n^3 \rho [v_p^2 \alpha_p^2 + 4(\omega_1 - \omega_{10})^2]} \tag{7.114}$$

where the approximation $k_1 \approx k_2$ and (7.100) have been used. It is evident that g_b will be reduced to one-half of its peak value when $\omega_1 - \omega_{10} = v_p \alpha_p/2$. The *Brillouin linewidth*, $\Delta\omega_b$, is defined as the full-width at half-maximum of the g_b function and is seen to be equal to $v_p \alpha_p$.

An important figure of merit in the Brillouin scattering process is the *threshold condition*, defined as the combination of parameters such that the gain for the Stokes wave over a given distance is equal to its loss over the same distance. Of particular interest is the input pump intensity needed to accomplish this. The condition is found from (7.113a) by setting $dI_1/dz = 0$ and $I_2 = I_2(0)$. The pump intensity at threshold is found to be

$$I_2(0)_T = \frac{\alpha}{g_b(\omega_1 = \omega_{10})} = \frac{2\epsilon_0 n^3 \rho \alpha v_p \Delta \omega_b}{\gamma^2 k_p k_2 \eta_0} \tag{7.115}$$

For the case in which there is weak nonlinear coupling, so that $I_1 \ll I_2$, the solutions of (7.113b) and (7.113a) are

$$I_2(z) = I_{20} \exp(-\alpha z) \tag{7.116}$$

$$I_1(z) = I_{10} \exp(\alpha z) \exp\left(-\frac{g_b I_{20}[1 - \exp(-\alpha z)]}{\alpha}\right) \tag{7.117}$$

where $I_{20} = I_2(z = 0)$ is the input pump intensity, and where $I_{10} = I_1(z = 0)$ is the output Stokes intensity. In the actual process, a Stokes input at $z = L$ would grow to produce an output intensity at $z = 0$. Considering I_{1L} as the given quantity, (7.117) yields the Stokes output:

$$I_{10} \approx I_{1L} \exp(-\alpha L) \exp\left(\frac{g_b I_{20}[1 - \exp(-\alpha L)]}{\alpha}\right) \tag{7.118}$$

Usually, no input at $z = L$ exists, and the Stokes wave builds up from noise resulting from initial pump scattering over a range of frequencies and directions. The stimulated process develops as scattered components near frequency ω_1 strengthen the acoustic wave at ω_p, and thus the phase-matched interaction eventually dominates. The SBS process in fibers, to be explored in Chapter 8, is most often to be avoided if possible. Therefore the threshold condition and its relation to the source and SBS linewidths are of particular interest.

PROBLEMS

7.1. Carry out the steps needed to derive the second order susceptibility, Eq. (7.10), from the anharmonic oscillator model, where the equation of motion is given by (7.9).

7.2. Perform the field product in Eq. (7.16) for the wave interaction shown in Fig. 7.2. Determine the resulting polarization terms, which include the appropriate frequencies and degeneracy factors.

7.3. Using appropriate geometrical constructions, verify Eq. (7.22). *Hint:* This can be done by considering a copolarized field interaction that occurs through the tensor element χ_{1111} in an isotropic medium. The fields and the nonlinear polarization (parallel to the fields) are next rotated through an arbitrary angle, thus enabling the rotated polarization to be described in terms of off-diagonal tensor elements. Requiring invariance between the rotated and nonrotated polarizations leads to the condition given by (7.22).

7.4. Consider an interaction in which two strong fields at frequency ω_a propagate x-polarized in the z direction. A third weak field at ω_b propagates collinearly with the other two, but it is y-polarized. Determine the nonlinear refractive index encountered by the weak wave in the presence of the two strong waves. How does the effective n_2' for this process compare to that found in Section 7.6?

7.5. Consider a gaussian pulse represented by the envelope

$$|E_0(t)|^2 = A \exp [-(t/\tau)^2]$$

Show that when undergoing self-phase modulation, the chirping introduced on the pulse is approximately linear in the vicinity of the envelope peak.

7.6. Show that (7.60) converts to (7.61) by means of the transformation $t_L = t - \beta_1 z$.

7.7. Verify that the hyperbolic secant pulse expressed in (7.62) is a solution to the nonlinear Schrödinger equation (7.61).

7.8. Determine the full-width at half-maximum of the pulse power in (7.62) in terms of the pulse width parameter, T.

7.9. Derive (7.57) by first differentiating (7.56) with respect to z; then express the resulting two terms as functions of appropriate time derivatives of (7.56). Equation (7.57) is constructed by using these identifications.

7.10. A *dark pulse* consists of a momentary drop in intensity from an otherwise uniform power level. Determine the basic requirements that would enable such a pulse to be a time-invariant solution to the nonlinear Schrödinger equation.

7.11. Consider the input to a medium as light at two frequencies, ω_1 and ω_2, having respective intensities I_{10} and I_{20}, and where $\omega_1 < \omega_2$. The light waves will be coupled through the Raman effect, as described by Eqs. (7.65) to (7.67). The medium is characterized by Raman gain coefficient g_r (for the two frequencies under consideration) and loss coefficient α. Suppose that losses are low, such that $\alpha z \ll 1$ for all realistic distances. Determine an expression for the distance at which the intensities of the two waves will equalize. What condition is placed on the relation between I_{10} and I_{20} for equalization to be possible?

7.12. Suppose two light waves at frequencies ω_1 and ω_2, where $\omega_1 < \omega_2$, are propagating in *opposite directions* in a medium in which stimulated Raman scattering can occur. Show that nonlinear coupling between these two waves is a phase-matched process, and thus the wave at ω_1 can experience gain in the presence of the wave at ω_2.

7.13. Verify that for stimulated Brillouin scattering in a lossless medium ($\alpha = 0$), the solutions of (7.113a) and (7.113b) are

$$I_1(z) = I_0 \frac{\psi_b}{1 - \psi_b}$$

$$I_2(z) = I_0 \frac{1}{1 - \psi_b}$$

where $I_0 = I_{20} - I_{10}$ and where $\psi_b = (I_{10}/I_{20}) \exp(-g_b I_0 z)$.

REFERENCES

1. A. Yariv, *Quantum Electronics*, 3rd ed. John Wiley & Sons, New York, 1989.

2. P. N. Butcher and D. Cotter, *The Elements of Nonlinear Optics*. Cambridge University Press, New York, 1990.

3. Y. R. Shen, *Nonlinear Optics*. John Wiley & Sons, New York, 1984, chap. 17.

4. D. Marcuse, "Selected Topics in the Theory of Telecommunications in Fibers," in *Optical Fiber Telecommunications II*, S. E. Miller and I. P. Kaminow, eds. Academic Press, Orlando, FL, 1988.

5. M. D. Levenson and S. S. Kano, *Introduction to Nonlinear Laser Spectroscopy*, rev. ed. Academic Press, Boston, 1988.

6. C. L. Tang, "Saturation and Spectral Characteristics of the Stokes Emission in the Stimulated Brillouin Process," *Journal of Applied Physics*, vol. 37, pp. 2945–2955, 1966.

8

Nonlinear Effects in Silica Fibers

Glass fibers for optical communications are made of fused silica, an amorphous material, to which dopant materials of various kinds are added to produce changes in refractive index. A number of third order nonlinear processes can occur; these can grow to appreciable magnitudes over the long lengths available in fibers, even though the nonlinear coefficients in the materials are relatively small. The effects are particularly important in single-mode fibers, in which the small mode field dimensions result in substantially high light intensities with relatively modest power inputs. Because of the long propagation distances usually required to achieve measurable nonlinear effects, only phase-matched processes are of any significance.

Nonlinear effects become increasingly important when light pulses are used, such as in a digitally coded signal. The high peak intensities of pulses within a pulse code modulated signal (compared to a cw input with the same average power) can result in significant enhancement of certain nonlinear effects that can severely degrade—or in some cases enhance—the signal. For example, stimulated Raman scattering and four-wave mixing can lead to cross-talk between signals in a wavelength division multiplexed system. Stimulated Brillouin scattering (producing backscatter) can be appreciable at input intensities on the order of a few milliwatts for extremely narrow band signals, but it becomes insignificant at powers below 100 W when short (broad band) pulses are used. Self- and cross-phase modulation (arising from the nonlinear change in refractive index) will lead to other pulse-shaping effects, which can greatly influence system performance as data rates approach 10 Gbits/s [1].

8.1. NONLINEAR PULSE PROPAGATION

Pulse propagation in fibers, in which dispersion and self-phase modulation occur, is governed by an appropriately modified form of the nonlinear Schrödinger equation (7.61). It is rewritten below:

$$\frac{\partial E_0(z, t_L)}{\partial z} = -\frac{\alpha}{2} E_0(z, t_L) + j \frac{\beta_2}{2} \frac{\partial^2 E_0(z, t_L)}{\partial t_L^2}$$
$$- j \frac{n_2'}{n_0} \beta_0 |E_0(z, t_L)|^2 E_0(z, t_L) \tag{8.1}$$

where $E_0(z, t_L)$ is the pulse electric field envelope function, and where $t_L = t - \beta_1 z$. t_L is the "local time" measured from the pulse peak, where the peak is at position z along the propagation path.

Equation (8.1) assumes the propagation of uniform plane waves, which is not the case in the fiber waveguide. If a single-mode fiber is assumed, the field amplitude in (8.1) must include the effect of the transverse field dependence, which is assumed gaussian:

$$E_0(r, z, t_L) = E_0(z, t_L) \exp\left(-r^2/r_0^2\right) \exp\left(-\alpha z/2\right) \qquad (8.2)$$

where r_0 is the mode field radius. Equation (8.2) is not substituted directly into (8.1), but it is used to determine an effective amplitude in terms of the pulse peak power, which is then used in (8.1). To accomplish this, $E_0(z, t_L)$ can be expressed as the product of a peak real amplitude, E_0, and a normalized field shape function, $f(z, t_L)$, that includes all of the shape and phase information in $E_0(z, t_L)$:

$$E_0(z, t_L) = E_0 f(z, t_L) \qquad (8.3)$$

Using the results of Problem 3.6, the power in the gaussian mode is then:

$$P(z, t_L) = \frac{n_0 A_{\text{eff}}}{4\eta_0} E_0^2 \exp\left(-\alpha z\right) |f(z, t_L)|^2 = P_0 \exp\left(-\alpha z\right) |f(z, t_L)|^2 \qquad (8.4)$$

where n_0 is taken to be either the core or the cladding index, and where the *effective area* is $A_{\text{eff}} = \pi r_0^2$. The peak power in the pulse is $P_0 = n_0 A_{\text{eff}} E_0^2 / 4\eta_0$. The complex field amplitude that appears in (8.1) can thus be written in terms of the peak power as

$$E_0(z, t_L) = \sqrt{\frac{P_0}{A_{\text{eff}}}} \exp\left(\frac{-\alpha z}{2}\right) u(z, t_L) \qquad (8.5)$$

where the *normalized power function*, $u(z, t_L)$, is related to the normalized field function through $u(z, t_L) = f(z, t_L) \sqrt{4\eta_0/n_0}$. Substituting (8.5) into (8.1) results in

$$\frac{\partial u(z, t_L)}{\partial z} = j \frac{\beta_2}{2} \frac{\partial^2 u(z, t_L)}{\partial t_L^2} - j \frac{n_2' \beta_0 P_0}{n_0 A_{\text{eff}}} \exp\left(-\alpha z\right) |u(z, t_L)|^2 u(z, t_L) \qquad (8.6)$$

The form of (8.6) is made simpler by incorporating additional normalized parameters. First, the local time, t_L, can be related to the inital pulse width,

T, by defining the normalized local time, $\tau_L = t_L/T$. Second, the propagation distance can be related to the *dispersion length*, L_D, by defining the normalized length parameter, $\xi = z/L_D$. The dispersion length is defined as

$$L_D = T^2/|\beta_2| \tag{8.7}$$

For a transform-limited gaussian pulse, L_D is the distance over which the pulse spread is equal to the initial pulse width, T, as was found in Problem 5.4. The above definition of L_D is in fact used for any pulse shape, although for shapes other than gaussian, the significance of the length may be different. Finally, the *nonlinear length* is defined as

$$L_{NL} = \frac{cA_{\mathrm{eff}}}{\omega_0 n_2' P_0} \tag{8.8}$$

where it is assumed that n_2' is positive (such is the case for silica). Recognizing that $\beta_0 = n_0\omega_0/c$ and using the above parameters, (8.7) becomes

$$\frac{\partial u(\xi,\tau_L)}{\partial \xi} = \pm j \frac{1}{2} \frac{\partial^2 u(\xi,\tau_L)}{\partial \tau_L^2} - jN^2 \exp(-\alpha z)|u(\xi,\tau_L)|^2 u(\xi,\tau_L) \tag{8.9}$$

where

$$N \equiv \left(\frac{L_D}{L_{NL}}\right)^{1/2} = \left(\frac{\omega_0 n_2' P_0 T^2}{|\beta_2| c A_{\mathrm{eff}}}\right)^{1/2} \tag{8.10}$$

The positive or negative sign in (8.9) is chosen on the basis of whether β_2 is positive or negative (or on the basis of D being negative or positive). The value of n_2' is 6.1×10^{-23} m^2/V^2 [2]. This is converted into units appropriate for use in (8.10) by multiplying the above n_2' value by the factor $2\eta_0/n_0$, where $n_0 = 1.46$ and $\eta_0 = 377$ ohms. Thus $n_2' \approx 3.2 \times 10^{-20}$ m^2/W.

Equation (8.9) is the most convenient form of the nonlinear Schrödinger equation, since it is completely normalized. The significance of N, the only variable parameter, is that it provides a measure of the relative strengths of the nonlinearity and the dispersion; large values of N, for example, indicate a relatively strong nonlinearity. It is a convenient measure, since it characterizes the interplay between the key parameters such as dispersion, pulse power, pulse width, and the nonlinear index, n_2'. Despite its simplified form, (8.9) must usually be solved numerically, using one of the methods described, for example, in [3]. Of concern in this study are the principal results of operation in regimes where the dispersion is negative or positive, as will be considered in the following sections.

8.2. OPTICAL SOLITONS

An analytic solution of (8.9) occurs when $N = 1$, and when the minus sign is chosen for the first term on the right-hand side. The solution is

$$u(\xi, \tau_L) = \operatorname{sech}(\tau_L) \exp(j\xi) \qquad (8.11)$$

The power function associated with the above solution, proportional to $|u|^2$, will be invariant with time and position. Consequently, (8.11) is the field function of an optical soliton. With $N = 1$, (8.10) with (8.7) and (8.8) yields

$$\frac{P_0}{A_{\text{eff}}} = \frac{c|\beta_2|}{\omega_0 n_2' T^2} \qquad (8.12)$$

This result is equivalent to the condition given by (7.63) for the solution of (7.61). The choice of the minus sign in (8.9) indicates that β_2 will be negative, or D is positive. Consequently, a soliton can exist only in the region of anomalous group velocity dispersion, at wavelengths greater than λ_0. This will result in the chirping introduced on the pulse by group dispersion canceling that produced by self-phase modulation, in the manner described in Section 7.7.

The pulse function given by (8.11) is known as the *fundamental soliton* ($N = 1$ case). Higher order solitons exist for integer values of N greater than 1. In these cases, the pulses no longer propagate without changing shape, but do so in a periodic fashion, reverting back to the original shape after a distance z_0, known as the *soliton period*. Simple analytic solutions for the higher order solitons do not exist, but their behavior can be determined through numerical solution of (8.9). Requiring N to be an integer leads to the establishment of higher order solitons by integer multiples of the product $T\sqrt{P_0}$, as (8.10) shows. Note that this product is proportional to *pulse area*, defined through

$$\text{Area} = \int_{-\infty}^{\infty} \sqrt{P_0} \operatorname{sech}\left(-\frac{t_L}{T}\right) dt_L = \pi T \sqrt{P_0} \qquad (8.13)$$

As a result of the uncertainty in the structure of higher order solitons at a given position, their applicability to communication systems is questionable. As of this writing, however, fundamental solitons are being considered for use in long-haul systems, such as undersea links.

The effect of loss on soliton propagation is to reduce the amplitude (and hence area) as the pulse propagates. Once the area is reduced below the required value, the pulse is subject to distortion by dispersion. The viability of solitons for use in systems is enhanced considerably by the availability of optical amplifiers that can be spaced periodically along the link to compensate fiber losses. Extremely low-loss fiber is still necessary from an economic standpoint, however, to allow large spacing between amplifying stations.

Finally, it is possible to achieve soliton behavior in the wavelength range less than λ_0 by the use of what are known as "dark pulses." These are momentary drops in intensity, rather than the usual rises in intensity that characterize a "bright" pulse. The effect is to reverse the sign of the chirping produced by self-phase modulation, leading to the possibility of having the nonlinear and normal linear dispersions cancel. Dark solitons have been a subject of active research [4] but their applicability to communication systems appears doubtful, chiefly because they exist in the spectral region where fibers exhibit relatively high loss.

8.3. OPTICAL PULSE COMPRESSION

Another application of self-phase modulation in fibers concerns pulses with frequency content in the normal dispersion regime ($\lambda < \lambda_0$). In this case, the linear and nonlinear dispersions are both positive, leading to substantial pulse broadening with chirping, accompanied by the added frequency content arising from self-phase modulation. The chirping is nearly linear over most of the pulse width. This combination of increased bandwidth and linear chirping makes the pulse amenable to compression after leaving the fiber, using a fairly simple apparatus [5]. Typically, a pair of gratings (Fig. 8.1) is used, which has the effect of introducing anomalous dispersion. The delay associated with the optical path through the grating pair can be made to vary almost linearly with wavelength by using a large spacing between gratings.

The results are impressive, in that pulses can typically be compressed by a factor of ten or more using a single stage (fiber followed by grating pair). In addition, multiple stages can be used. In another technique the light pulse propagates through a sequence of prisms which introduces anomalous dispersion and provides correction for cubic dispersion [6]. The record for the shortest

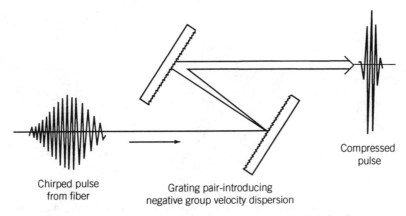

Compressed
pulse

Chirped pulse
from fiber

Grating pair-introducing
negative group velocity dispersion

Figure 8.1. Grating pair apparatus used to compress a chirped pulse.

pulse ever generated (6 fs) was established using this method [7]. Commercial pulse compressors designed for use with specific laser systems are available.

8.4. STIMULATED RAMAN SCATTERING

Stimulated Raman–Stokes scattering is likely to occur in fibers, producing appreciable amplification for down-shifted signals when moderate pump power levels are used. As demonstrated in Chapter 7, this is largely because the process is phase matched, so that the long interaction lengths available in fibers will enhance the process. Additionally, the small mode field cross section yields high intensities for relatively modest power levels. Under these conditions, the generation of Stokes frequencies will usually begin with spontaneous Raman scattering, with the input of light at one frequency. As the scattered light grows in intensity, the process becomes stimulated, which can lead to Stokes power levels that are comparable to that of the pump input. To describe the effect, the theory of Chapter 7 must be modified to include the effect of spontaneous scattering.

The Raman effect in fibers arises from a few basic vibrational modes that occur in fused silica and in structurally related glasses that are used as dopants. As described in Chapter 4, the structural model for amorphous silica consists of a continuous network of SiO_4 tetrahedra. The corner oxygen atoms in each tetrahedron are shared by adjacent units, thus maintaining the SiO_2 formula. The model is based on x-ray studies and fundamental bonding arguments. Several vibrational modes occur in this structure that give rise to Raman scattering. Among these are the "rigid cage" mode (Fig. 8.2a) in which the four oxygen atoms move as a unit in one direction, while the silicon atom moves in the opposite direction. This mode gives rise to the Raman frequency of 1065 cm^{-1}. Another is the "breathing" mode (Fig. 8.2b) in which the oxygen atoms move symmetrically to expand and contract the tetrahedron, with the silicon atom stationary, resulting in the 800 cm^{-1} Raman line. The strongest Raman frequencies are at 440 cm^{-1} and 490 cm^{-1}. These arise from a more complicated mode that is depicted in Fig. 8.2c. The transverse optical (TO) form of this mode gives

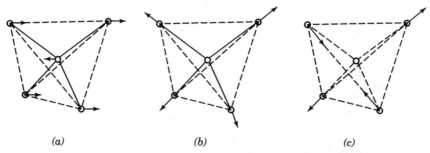

(a)	*(b)*	*(c)*

Figure 8.2.   Raman oscillation modes in fused silica: (a) 1056 cm^{-1}, (b) 800 cm^{-1}, and (c) 440 cm^{-1} [8].

rise to the 440 cm^{-1} line; the 490 cm^{-1} line is the longitudinal optical (LO) version of the mode, which includes the effects of long-range Coulomb forces [9]. The presence of other lines in addition to broadening effects produce the net Raman gain spectrum as shown in Fig. 8.3a.

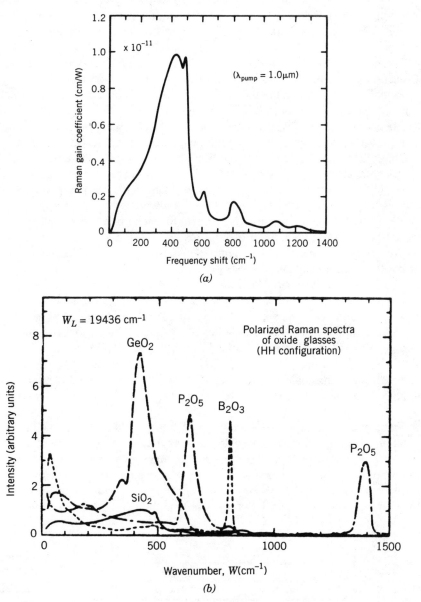

Figure 8.3. (a) Raman gain spectrum in fused quartz (SiO$_2$) for which the pump wavelength is $\lambda_2 = 1.0\,\mu$m. (Adapted from and from ref. 11, adapted from ref. 10) © 1980 IEEE. (b) Raman spectra of component glasses used in fiber manufacture [12].

Other glasses used as dopants of pure silica in fiber manufacture exhibit Raman gain that can be significantly higher than that of silica. These materials, considered in Chapter 4, include GeO_2, B_2O_3, and P_2O_5. Their Raman spectra are shown in Fig. 8.3b, along with that of silica. Use of the dopant materials can as a rule cause significant enhancement of SRS over that in pure silica. When pumped by light at 514.5 nm wavelength, the peak Raman cross sections exhibited by pure GeO_2, B_2O_3, and P_2O_5 normalized to that of silica are 8.2, 4.7, and 5.7 [12].

The curve in Fig. 8.3a shows the gain experienced by the Stokes wave as a function of its frequency shift *below* that of the pump frequency (occurring at 0 cm^{-1}). The width of the gain curve (FWHM) is approximately 270 cm^{-1}, which corresponds to about 8 THz, where $f(Hz) = cf(cm^{-1})$. Optical signals whose bandwidths are on this order or less are thus susceptible to considerable amplification in the presence of a pump wave of appropriately displaced wavelength. In addition, the gain for the Stokes wave decreases with increasing pump wavelength, as can be inferred from (7.84). Thus, in general, $g_r = (\lambda_d/\lambda_2)g_r|_{\lambda_d}$, where λ_2 is the pump wavelength to be used and λ_d is the pump wavelength (in this case 1.0 μm) at which the data are known.

Two theoretical models of SRS in fibers that incorporate spontaneous scattering include one for the case of high losses [13] and a more general theory for low-loss fibers that includes pump depletion effects [14]. The main results of the latter theory are presented here.

The interaction geometry is shown in Fig. 8.4. The single-mode fiber is of length L. Inputs at both the Stokes and pump wavelengths are assumed, having powers P_{10} and P_{20}, respectively. The pump power is at a single frequency, ω_2; the Stokes input power is distributed over a bandwidth $\Delta\omega_1$ that is centered at frequency ω_1. Raman gain is assumed to occur over a spectrum of Lorentzian shape having linewidth $\Delta\omega_r$; $\Delta\omega_1$ and $\Delta\omega_r$ are not necessarily equal. Loss arising from the mechanisms considered in Chapter 4 is assumed to be the same at both wavelengths and is characterized by exponential power loss coefficient α.

Since gain occurs for Stokes power over the entire Raman bandwidth, computation of the net Stokes power at the fiber output necessitates that an integral of power as a function of frequency be taken over the Raman lineshape, the details of which are presented in [14]. Rather than treating the frequency-dependent power as a continuous function, it is more convenient to consider the power as appearing in a set of discrete longitudinal modes that lie within an

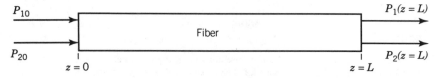

Figure 8.4. Beam geometry for SRS in a single-mode fiber.

appropriately modified Raman gain spectrum. A longitudinal mode is defined as a frequency at which light enters and exits the fiber with the same phase. The frequency spacing between adjacent modes is thus $\Delta f = c/n_{eff} L$, where n_{eff} is the effective index of the LP_{01} mode.

Using the discrete mode picture, the total Stokes power is found by counting the number of longitudinal modes, q, that fall within an effective Raman bandwidth, B_{eff}, inside of which the gain is a constant; outside B_{eff} the gain is zero. The form of B_{eff} arises from requiring equivalence between the continuous spectrum power calculation using the true lineshape and the calculation resulting from the discrete mode model. The result is [14]

$$B_{eff} = \frac{\Delta\omega_r}{4\sqrt{\pi}} \left(\frac{g_r P_{20} L_{eff}}{A_{eff}} + W_r \right)^{-1/2} = \frac{qc}{n_{eff} L} \qquad (8.14)$$

where $W_r = (\Delta\omega_r/\Delta\omega_1)^2$ when $P_{10} \neq 0$; when $P_{10} = 0$, then $W_r = 0$. A_{eff} is the effective area of the fiber mode, equal to πr_0^2, where r_0 is the mode field radius. g_r is the Raman power gain coefficient at frequency ω_1, expressed in cm/W. g_r is given by (7.84) or is experimentally measured. The *effective length* parameter, L_{eff}, is defined in terms of the exponential power loss coefficient α and the fiber length L as

$$L_{eff} \equiv \int_0^L \exp(-\alpha z)\, dz = \frac{1 - \exp(-\alpha L)}{\alpha} \qquad (8.15)$$

Note that for low-loss situations, $L_{eff} \approx L$, whereas for high loss, $L_{eff} \approx 1/\alpha$.

An important feature of B_{eff} is the fact that it decreases as pump power is increased. The logic of this is seen by considering a Lorentzian gain profile (assuming a single Raman resonance) whose amplitude increases with pump power. Consider taking the ratio of the amplified power of a mode having frequency near the line center and that of one which is tuned far from the line center. Since the gain factor is an exponential function, this ratio will increase as the Lorentzian amplitude increases. As a result, the off-resonance modes carry a decreasing percentage of the total Stokes power as the pump power is raised; the effective bandwidth is thus reduced.

The pump and Stokes intensities are governed by the differential equations, (7.83a) and (7.83b). The solutions will again be (7.85) and (7.86), which can be written in terms of the signal powers, where in general $P = I A_{eff}$. At length L, the solutions become

$$P_1(L) = \frac{\omega_1}{\omega_2} P_0 \exp(-\alpha L) \left(\frac{\psi_r}{1 + \psi_r} \right) \qquad (8.16)$$

$$P_2(L) = \frac{P_0 \exp(-\alpha L)}{1 + \psi_r} \tag{8.17}$$

where $P_0 = P_{20} + (\omega_1/\omega_2)P_{10}$. ψ_r, the coupling parameter first defined in (7.87), now includes the spontaneous scattering effect through [14]

$$\psi_r = \frac{\omega_2}{\omega_1}\left(\frac{\hbar\omega_1 B_{\text{eff}}}{P_0} + \frac{P_{10}}{P_{20}}\right)\exp(G_0) \tag{8.18}$$

where G_0 is a gain parameter given by

$$G_0 = \frac{g_r P_0 L_{\text{eff}}}{A_{\text{eff}}} \tag{8.19}$$

The effect of ψ_r is understood by observing that if $\psi_r \ll 1$, then (8.16) and (8.17) show that negligible conversion occurs from the pump to the Stokes wave. If, on the other hand, $\psi_r \gg 1$, then the equations show that complete power transfer from pump to Stokes will occur.

For the case in which $P_{10} = 0$, stimulated Raman scattering will build up from spontaneously emitted Stokes photons. In this case (8.18) becomes, using (8.14),

$$\psi_r = \left(\frac{q\hbar\omega_2 c}{n_{\text{eff}} L}\right)\frac{1}{P_{20}}\exp(G_2) \tag{8.20}$$

where the gain parameter in the absence of Stokes input is $G_2 = g_r P_{20} L_{\text{eff}}/A_{\text{eff}}$. Using (8.20) in (8.16), assuming $\psi_r \ll 1$, the Stokes power becomes

$$P_1(L) = \left(\frac{q\hbar\omega_1 c}{n_{\text{eff}} L}\right)\exp(-\alpha L)\exp(G_2) \tag{8.21}$$

An interesting interpretation of (8.21) is that the first bracketed term in the equation is the power in watts of an input of one Stokes photon (having an energy of $\hbar\omega_1$) per longitudinal mode. This constitutes the effective input power, P_{10}, for the case in which the Stokes wave initiates from spontaneous scattering alone. Using (8.14) again, (8.20) can be expressed in a form that is more amenable to computation using measured parameters:

$$\psi_r = \frac{\hbar\omega_2\Delta\omega_r}{4\sqrt{\pi}} \frac{1}{P_{20}} G_2^{-1/2} \exp(G_2) \tag{8.22}$$

The case in which $\psi_r = 1$ defines the *critical condition*, at which $P_1(L) = P_2(L)$ as (8.16) and (8.17) show.* The value of P_{20} for this case is typically of interest for given values of wavelength, fiber length, loss, and beam cross section. One method of solving (8.22) for P_{20} is to use the result that G_2 is usually in the vicinity of 16 at the critical condition. This trial value of G_2 can be substituted into (8.22) with $\psi_r = 1$, followed by slight iterative modifications of P_{20} until the equation is satisfied. The $G_2 \approx 16$ condition is in fact a widely used estimation method based on a previous theory that assumes fibers of high loss and does not account for pump depletion [13].

The above discussion assumes all fields are linearly polarized in the same direction, and that the input and Stokes powers are continuous. In the case of random polarization (which would occur in a round fiber), estimates of P_{20} at the critical condition should be approximately doubled. This is because Raman gain is zero when pump and Stokes fields are orthogonally polarized.

When pulses are used, the Raman interaction will decrease since the pump and Stokes pulses separate as they propagate at different group velocities. Energy transfer between the two pulses will occur only when they spatially overlap. In the case of a single input pump pulse, the generated Stokes pulse will separate from the pump, given sufficient distance. This phenomenon, known as *walkoff*, has the effect of reducing the interaction length, L_{eff}, to the propagation distance over which one pulse passes through the other [15]:

$$L_w = 2T \left| \frac{v_{gs}v_{gp}}{v_{gs} - v_{gp}} \right| \tag{8.23}$$

v_{gp} and v_{gs} are the group velocities of the pump and Stokes pulses, respectively, and T is the halfwidth of the pump pulse. In predicting Stokes power levels, good results have been obtained by using the walkoff length, L_w, in place of L_{eff} to determine G_2 in (8.21); the computations are then performed as before. The pulses are able to completely separate because, upon generating the Stokes pulse, the pump pulse energy is depleted by the interaction, thus effectively halting the nonlinear process.

SRS in fibers is further complicated by the occurrence of multiple orders. Specifically, Raman-shifted light from a primary interaction may be of sufficient strength to act as a pump source itself, resulting in the generation of an additional Stokes wave. Figure 8.5 shows an example of a fiber output spectrum in which multiple Raman orders occur [16]. The spectral broadening evident in the higher orders results from the effects of self-phase modulation.

---

*The critical condition is often referred to as the threshold condition for SRS in fibers.

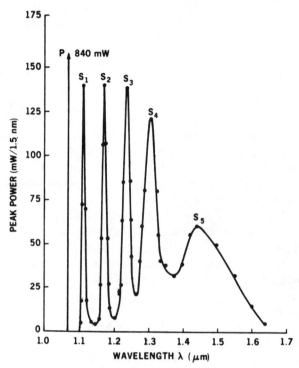

Figure 8.5. Output spectrum from a single-mode fiber Raman laser, showing multiple Stokes orders. Pump wavelength is $\lambda = 1.06\,\mu$m [16]. © 1978 IEEE.

Since SRS can be an efficient mechanism for frequency conversion, its presence is undesirable in communication systems. In a single-wavelength channel, frequency conversion can result in multiple copies of the transmitted signal, all of which will have different group delays. In systems that employ wavelength division multiplexing, SRS is a mechanism by which cross-talk may occur between signals whose wavelength separation falls within the Raman gain curve. Specifically, the longer wavelength signals will grow at the expense of those at shorter wavelengths, leading to power penalties for the latter signals. Intuitively, one would suspect that the severity of this effect would depend on various channel parameters such as their frequency spacing, the number of channels, and the power in each. Guidelines are thus needed that set limits on the above quantities such that cross-talk effects are minimized.

An analysis that determined the relationships between the above parameters was performed in [17]. In it, a link having N channels with equal frequency spacing, Δf, and equal powers is assumed. Since Raman frequency conversion occurs primarily from short to long wavelengths, the shortest wavelength channel will suffer the most degradation and will thus determine the limits that are to be set. Consider the interaction between two channels having input powers P_{10} (Stokes) and P_{20} (pump). Since inputs at both frequencies are present, spontaneous scattering will be negligible. Equation (8.18) thus becomes

$$\psi_r = \frac{\omega_2 P_{10}}{\omega_1 P_{20}} \exp(G_0) \approx \frac{\lambda_1}{\lambda_2} \frac{P_{10}}{P_{20}} (1 + G_0) \tag{8.24}$$

where it is assumed that G_0, given by (8.19), is small (as would be desirable). Now, assuming low loss and small ψ_r, (8.17) becomes $P_2(L) = P_{20}/(1 + \psi_r) \approx P_{20}(1 - \psi_r)$. The fraction of power lost by the pump over propagation distance L is thus

$$F_2 = \frac{P_{20} - P_2(L)}{P_{20}} \approx \frac{g_r P_{10} L_{\text{eff}}}{A_{\text{eff}}} \tag{8.25}$$

where it is assumed that $\lambda_1 \approx \lambda_2$, for purposes of evaluating their ratio.

The fractional power lost by the pump (of wavelength λ_p) to $N - 1$ channels of longer wavelength will be

$$F_N = \sum_{i=1}^{N-1} \frac{g_{ri} P_{i0} L_{\text{eff}}}{2 A_{\text{eff}}} \tag{8.26}$$

where L_{eff} and A_{eff} are assumed the same for all wavelengths. It is further assumed that Stokes conversion takes place only between the shortest wavelength channel and the others, with no power conversion between the longer wavelength channels. This assumption is appropriate for determining the worst-case situation of loss from the shortest wavelength signal. The factor of $\frac{1}{2}$ in (8.26) accounts for the effects of random polarization, in which the gain is effectively reduced by that factor. The gain, g_{ri}, will depend on the frequency separation between the pump and a given channel, as determined by the gain curve of Fig. 8.3a. To facilitate the evaluation of (8.26), the gain curve is modeled as approximately triangular, with zero gain at $\Delta f = 0$; the maximum value, g_{rp}, occurs at $\Delta f \approx 500$ cm$^{-1} = 1.5 \times 10^4$ GHz, with zero gain for Δf beyond 500 cm^{-1}. The Raman gain as a function of frequency displacement is thus

$$g_{ri} \approx \frac{\Delta f g_{rp}}{1.5 \times 10^4} \qquad \Delta f < 1.5 \times 10^4 \text{ GHz} \tag{8.27}$$

Further assumptions are as follows: (1) the channels have equal frequency spacing, Δf; (2) each channel carries the same power, P_{s0}; and (3) all channels fall within the Raman gain bandwidth. Then, using (8.27), (8.26) becomes [17]

$$F_N \approx \frac{\Delta f g_{rp}}{1.5 \times 10^4} \frac{P_{s0}L_{eff}}{2A_{eff}} \sum_{i=1}^{N-1} i = \frac{[(N-1)\Delta f][NP_{s0}]}{6 \times 10^4} \frac{g_{rp}L_{eff}}{A_{eff}} \qquad (8.28)$$

where Δf is in GHz. The terms in (8.28) are grouped to illustrate the effect on signal degradation produced by the total occupied bandwidth, $(N-1)\Delta f$, and the total optical power input, NP_{s0}. The equation demonstrates, for example, that an increase in the number of channels within a specified total bandwidth requires a corresponding decrease in the power per channel, such that signal degradation (measured by F_N) does not exceed a set value.

8.5. STIMULATED BRILLOUIN SCATTERING

As described in Chapter 7, Brillouin scattering involves the backscattering of light from a moving index grating formed by an acoustic wave that copropagates with the input light. The scattered light frequency is Doppler shifted from that of the input, so that $\omega_1 = \omega_2 - \omega_p$, where ω_1 and ω_2 are the backscattered (Stokes) and input (pump) frequencies, and where ω_p is the acoustic frequency. In the stimulated process, the acoustic wave is reinforced by the interaction of the pump and Stokes waves. Furthermore, as a result of the phase-matching requirements, the Stokes output is restricted to a frequency range determined by the Brillouin linewidth, $\Delta\omega_b = v_p\alpha_p$, where v_p and α_p are, respectively, the velocity and attenuation coefficient of the acoustic wave.

The treatment of stimulated Brillouin scattering in fibers proceeds in a similar manner to that used for stimulated Raman scattering. The optical field intensities are evaluated as the LP_{01} mode power divided by the mode effective area, A_{eff}. In the case of a nondepleted pump wave, (7.118) applies and assumes the form

$$P_{10} \approx P_{1L} \exp(-\alpha L) \exp\left(\frac{g_b P_{20} L_{eff}}{A_{eff}}\right) \qquad (8.29)$$

where P_{10} and P_{20} are the output Stokes and input pump powers, and where g_b is the Brillouin gain, defined by (7.114). The effective length, L_{eff}, is defined by (8.15) as before.

In a treatment analogous to that of Raman scattering, the Stokes input, P_{1L}, is treated as an effective quantity that is equivalent to the power contained in q Stokes photons, where q is the number of longitudinal modes within the effective bandwidth given by (8.14) with $W_r = 0$ [13]. The analysis in [13] leads to a critical condition, analogous to that found from (8.22) for Raman scattering, at which $P_{10} = P_{20}$:

$$\frac{\omega_1 k_B T \Delta \omega_b}{4\sqrt{\pi} \omega_p P_{20}} G_b^{-3/2} \exp G_b = 1 \tag{8.30}$$

where k_B is Boltzmann's constant, T is the temperature in kelvins, and where

$$G_b = \frac{g_b P_{20} L_{\text{eff}}}{A_{\text{eff}}} \tag{8.31}$$

Numerical values for the parameters include $g_b = 4.5 \times 10^{-9}$ cm/W at line center [18], $\Delta \omega_b \approx 10^9$ s^{-1}, and the value of $k_B T/\hbar \omega_p$ between 100 and 200 [13]. Using these values, a procedure similar to that used in analyzing (8.22) yields the approximation $G_b \approx 21$ [13]. This satisfies (8.30) with reasonable accuracy for most cases.

The peak Brillouin gain specified above is seen to be over two orders of magnitude larger than the peak Raman gain ($g_r = 7 \times 10^{-12}$ cm/W at 1.55 μm). In spite of this, the Stokes power arising from Brillouin scattering can be less than that for Raman scattering in situations where the pump bandwidth greatly exceeds the Brillouin linewidth. Specifically, the effective gain for Brillouin scattering is of the form [18]

$$g_b(\Delta \omega) \approx g_b \frac{\Delta \omega_b}{\Delta \omega} \tag{8.32}$$

where $\Delta \omega$ is the pump bandwidth. An additional reduction by a factor of $\frac{1}{2}$ occurs for the case of random polarizations of the pump and Stokes waves, since the interaction requires copolarized fields. With Brillouin linewidths on the order of 100 MHz, it is evident that in communication channels employing modulation rates of several GHz, the SBS effects are significantly reduced. In contrast, the Raman linewidth, being several THz, will support substantial forward scattering in these situations. A detailed analysis of the systems implications of SBS is presented in [19].

8.6. FIBER AMPLIFIERS

An important advance in optical fiber technology occurred with the development of fibers that amplify light through stimulated emission. Such fibers have been made by incorporating various rare-earth dopants into the core material, the most successful of which has been erbium (Er). The result for the Er-doped case is fiber that will provide gain at 1.55 μm when the fiber is "pumped" by additional light input at any one of a number of shorter wavelengths. Lengths of amplifying fiber are intended for use as repeater sections in communications systems, replacing the electronic units that are commonly used. The primary reason for the dramatic simplification introduced in a fiber amplifier repeater is that the transmitted signal remains in optical form throughout the link, rather than being transformed to an electrical signal and back to optical whenever an electronic repeater stage is encountered. In addition, a single fiber amplifier can be used for all channels in a wavelength division multiplexed signal, whereas a separate electronic repeater would be needed for each wavelength. Aside from systems applications, numerous device applications for signal processing, in addition to the construction of Er-doped fiber-based lasers have been proposed or demonstrated.

The mechanism of amplification by stimulated emission can be demonstrated using a simple material model. The material consists of N_t identical atoms per unit volume; each atom has four possible energy states associated with, for example, four possible electron configurations. Upward transitions between energy levels in a single atom occur through the absorption of an incident photon. In the absence of additional light, downward transitions occur either by nonradiative relaxation or by radiative relaxation through spontaneous emission (the random emission of a single photon in any direction). When an additional photon is incident on the atom, a downward transition can be stimulated, resulting in the emission by the atom of a second photon that propagates with the incident one. As more photons are generated, these in turn stimulate downward transitions in adjacent atoms; this cascading effect can ultimately result in substantial power gain, provided a sufficient number of atoms can initially be excited to the higher energy states and provided the number of downward transitions per unit time can be made to exceed the upward transition rate.

In the four-level model, light inputs at two different frequencies are separately responsible for absorption and emission (Fig. 8.6). The light that is absorbed, known as the pump energy, is at the higher frequency; its presence induces transitions from level 1 to level 4, whose energy difference is $\hbar\omega_2$. The pump light is input at frequency ω_2 to coincide with this resonance. Fast nonradiative transitions occur from level 4 to level 3, thus allowing a substantial number of atoms to assume the level 3 energy state. This buildup of "population" in level 3 is assured if relaxation processes from level 3 to level 2 or 1 are either slow or are not allowed.

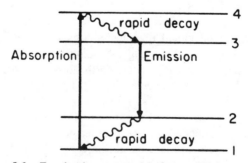

Figure 8.6. Four-level system model of an amplifying material.

The energy spacing between levels 3 and 2 is $\hbar\omega_1$. The model assumes that relaxation of population from level 3 to level 2 can occur through stimulated or spontaneous emission; the latter occurs with characteristic relaxation time τ. The $3 \rightarrow 1$ transition is assumed forbidden. From level 2, fast nonradiative decay again occurs to level 1. Since levels 4 and 2 both relax quickly, the populations of these levels are both essentially zero, meaning that the total atomic population, N_T, is divided in some proportion between levels 3 and 1; thus $N_T = N_1 + N_3$. This also means that the level 3 population can exceed that of level 2, resulting in a *population inversion* between these two levels. The result is that net gain can occur for light at frequency ω_1, since the rate of stimulated downward transitions between 3 and 2 will exceed the upward transition rate *between these levels*.

To determine the gain as a function of the various input and medium parameters, *rate equations* for the population densities of the important energy levels must be solved. Consider a fiber whose core is doped with erbium or another substance for which the excitation dynamics can be described by the four-level model. The rate equations that describe the populations of levels 1 and 3 are

$$\frac{dN_1}{dt} = -\frac{\sigma_{ap}A_p(V)}{A_c}\frac{P_p}{\hbar\omega_2}N_1 + \frac{\sigma_{es}A_s(V)}{A_c}\frac{P_s}{\hbar\omega_1}N_3 + \frac{N_3}{\tau} \quad (8.33)$$

$$\frac{dN_3}{dt} = -\frac{dN_1}{dt} \quad (8.34)$$

The pump and signal powers, P_p and P_s, are expressed in terms of photon flux densities (photons/s-m^2) by dividing both quantities by the energy per photon, $\hbar\omega_2$ or $\hbar\omega_1$, and the fiber core area, A_c (erbium is assumed to be present in the core only). σ_{ap} and σ_{es} are the absorption and emission crossections (expressed in m^2); these, when multiplied by the appropriate photon flux densities, would yield the probability of excitation or deexcitation of a single atom in a specified time period. Multiplying the crossections by the associated number densities of the ground or excited states (N_1 or N_3) would yield the exponential absorption or gain coefficients for the pump and signal waves. The expressions $A_p(V)$ and $A_s(V)$ are the fractions of the pump and signal powers that reside in the core (where absorption and gain exist), as given by (5.35).

Equations (8.33) and (8.34) are most easily solved in steady state, in which all time derivatives are zero. Resulting are expressions for N_1 and N_3 in terms of N_t:

$$N_1 = \frac{(1 + P_s/P_{\text{sat}}^{es})N_T}{(1 + P_p/P_{\text{sat}}^{ap} + P_s/P_{\text{sat}}^{es})} \quad (8.35)$$

$$N_3 = \frac{P_p/P_{\text{sat}}^{ap}N_T}{(1 + P_p/P_{\text{sat}}^{ap} + P_s/P_{\text{sat}}^{es})} \quad (8.36)$$

The saturation powers for the pump and signal are defined as

$$P_{\text{sat}}^{ap} = \frac{A_c \hbar \omega_2}{\sigma_{ap} A_p(V)\tau} \quad \text{and} \quad P_{\text{sat}}^{es} = \frac{A_c \hbar \omega_1}{\sigma_{es} A_s(V)\tau} \tag{8.37}$$

Note, for example, that in the absence of signal power, P_{sat}^{ap} is the pump power required to equalize the two populations.

The signal and pump powers grow or attenuate with distance in the fiber according to the equations

$$\frac{dP_p}{dz} = -\sigma_{ap} N_1 A_p(V) P_p \tag{8.38}$$

$$\frac{dP_s}{dz} = \sigma_{es} N_3 A_s(V) P_s \tag{8.39}$$

The equations are coupled since N_1 and N_3 both depend on P_p and P_s. An analytic solution can be obtained for the special case in which $P_p \ll P_{\text{sat}}^{ap}$ and $P_s \ll P_{\text{sat}}^{es}$. As a result, $N_1 \approx N_t$ and $N_3 \approx N_t(P_p/P_{\text{sat}}^{ap})$. Under these conditions, (8.38) is readily solved to yield

$$P_p(z) \approx P_p(0) \exp\left[-\sigma_{ap} N_t A_p(V) z\right] \tag{8.40}$$

Equation (8.39) can then be solved by assuming weak absorption for the pump, such that $P_p(z) \approx P_p(0)$. Then, using the above approximation for N_3, the result is

$$P_s(z) \approx P_s(0) \exp\left(\sigma_{es} A_s(V) N_t \frac{P_p}{P_{\text{sat}}^{ap}} z\right) \tag{8.41}$$

This result, although greatly simplified, demonstrates that at a given pump power level, the available gain will be appreciable if (1) the absorption and emission cross sections are high and (2) the lifetime of the metastable state (level 3) is long.

An energy level diagram for erbium is shown in Fig. 8.7. In this system, each indicated state is in fact a manifold of Stark levels, which at thermal equilibrium are populated according to a Boltzmann distribution. The latter gives the ratio of populations of any two Stark levels within a manifold as

$$\frac{N_a}{N_b} = \exp\left(\frac{(E_b - E_a)}{k_B T}\right) \tag{8.42}$$

where E_a and E_b are the energies associated with levels a and b, k_B is Boltz-

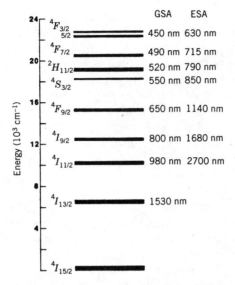

Figure 8.7. Energy level diagram for erbium, showing wavelengths for excitation from the ground state (GSA) and from the first excited state (ESA) [21]. © 1991 IEEE.

mann's constant, and T is the temperature in kelvins. Consequently, the lower levels of each manifold are the most populated. The contrast between populations of the lowest and highest levels in each state will be high if the manifold width is larger than k_BT. In the erbium system at room temperature, the manifold widths are between 300 and 400 cm^{-1}, whereas k_BT is approximately 200 cm^{-1}. In the diagram, manifolds are labeled using the standard spectroscopic notation, $^{(2S+1)}L_J$, where L (in letter code), S, and J, are the orbital, spin, and total angular momenta, respectively [20].

The Boltzmann distribution of population among the component levels in each manifold makes possible the formation of a population inversion. In the presence of pumping energy, population can be raised from the ground state ($^4I_{15/2}$) to the first excited state ($^4I_{13/2}$). There, the population quickly thermalizes to assume a Boltzmann distribution within that manifold, such that the lower levels of $^4I_{13/2}$ are densely populated. The $^4I_{13/2}$ state is metastable, and thus a population inversion can build up between its lower levels and the sparsely populated upper levels of the ground state. Transitions between $^4I_{13/2}$ and $^4I_{15/2}$ then occur either through spontaneous emission (observable as fluorescence) or by stimulated emission. Assuming the thermalization time within each manifold is much less than the lifetime of the manifold, the absorption and emission crossections as functions of frequency can be related through [21,22]

$$\sigma_{es}(\omega) = \sigma_{ap}(\omega)\exp\left[(E - \hbar\omega)/kT\right] \tag{8.43}$$

where E is defined as the photon energy at which the two crossections are equal; at higher energies (frequencies), absorption dominates, whereas at lower energies, the reverse is true. Also evident is that the separation between the two spectra will increase as temperature decreases.

Experimental plots of the absorption and emission crossections for the $^4I_{13/2} \rightarrow {}^4I_{15/2}$ transition in erbium, which exhibit the above behavior, are shown in Fig. 8.8. The fact that the emission crossection exceeds that of absorption at wavelengths beyond 1.54 μm explains the attraction of erbium in fibers for use in 1.55 μm wavelength systems. It is also seen that the possibility for pumping at shorter wavelengths occurs since absorption dominates emission in this range. Specifically, pump light at wavelengths in the vicinity of 1.50 μm can be used to boost population from the lower levels of $^4I_{15/2}$ to the upper levels of $^4I_{13/2}$. The thermalization process within the excited state gives rise to the shift in the emission spectrum from the absorption, thus allowing net gain at the longer wavelengths. In this scheme, light within the wavelength range between 1.45 and 1.50 μm typically serves as the pump, to induce gain for light near 1.55 μm. The system thus behaves in a qualitative way like the four-level system model described earlier. Use of pump and signal wavelengths that lie within the same transition in this way is termed *resonant pumping*.

The overlap of absorption and emission spectra is in fact an undesirable characteristic because (1) the pump energy acts to simultaneously deplete the upper level while populating it, thus limiting the obtainable inversion even at very

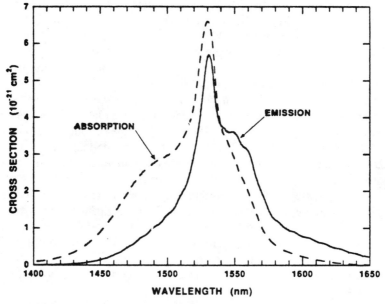

Figure 8.8. Absorption and emission spectra for erbium in Al/P silica (with aluminum and phosphorus co-doping) [21]. © 1991 IEEE.

high pump powers; and (2) the signal experiences absorption from the $1 \rightarrow 4$ transition while experiencing gain from $3 \rightarrow 2$. These effects can be included in the rate equations, (8.33) and (8.34), by incorporating two additional crosssections, σ_{as} and σ_{ep}. These are, respectively, the absorption crosssection at the signal wavelength and the emission crosssection at the pump wavelength. With these added parameters, (8.33) becomes

$$\frac{dN_1}{dt} = \frac{(\sigma_{ep}N_3 - \sigma_{ap}N_1)A_p(V)}{A_c} \frac{P_p}{\hbar\omega_2}$$
$$+ \frac{(\sigma_{es}N_3 - \sigma_{as}N_1)A_s(V)}{A_c} \frac{P_s}{\hbar\omega_1} + \frac{N_3}{\tau} \tag{8.44}$$

Again, with $N_3 = N_t - N_1$, the steady-state expressions for N_1 and N_3 are found from (8.44):

$$N_1 = \frac{(1 + P_s/P_{\text{sat}}^{es} + P_p/P_{\text{sat}}^{ep})N_t}{[1 + P_p(P_{\text{sat}}^{ep} + P_{\text{sat}}^{ap})/(P_{\text{sat}}^{ep}P_{\text{sat}}^{ap}) + P_s(P_{\text{sat}}^{es} + P_{\text{sat}}^{as})/(P_{\text{sat}}^{es}P_{\text{sat}}^{as})]} \tag{8.45}$$

$$N_3 = \frac{(P_s/P_{\text{sat}}^{as} + P_p/P_{\text{sat}}^{ap})N_t}{[1 + P_p(P_{\text{sat}}^{ep} + P_{\text{sat}}^{ap})/(P_{\text{sat}}^{ep}P_{\text{sat}}^{ap}) + P_s(P_{\text{sat}}^{es} + P_{\text{sat}}^{as})/(P_{\text{sat}}^{es}P_{\text{sat}}^{as})]} \tag{8.46}$$

where the saturation powers corresponding to the added crosssections are

$$P_{\text{sat}}^{as} = \frac{A_c\hbar\omega_1}{\sigma_{as}A_s(V)\tau} \quad \text{and} \quad P_{\text{sat}}^{ep} = \frac{A_c\hbar\omega_2}{\sigma_{ep}A_p(V)\tau} \tag{8.47}$$

In this new model, (8.38) and (8.39) become

$$\frac{dP_p}{dz} = (\sigma_{ep}N_3 - \sigma_{ap}N_1)A_p(V)P_p \tag{8.48}$$

$$\frac{dP_s}{dz} = (\sigma_{es}N_3 - \sigma_{as}N_1)A_s(V)P_s \tag{8.49}$$

To see qualitatively the effect of spectral overlap on the signal, (8.49) can be solved under the condition in which both power levels are much less than any saturation power value. Assuming in addition that $P_s \ll P_p$, the result is (Problem 8.6)

$$P_s \approx P_s(0)\exp\left[\left(\sigma_{es}\frac{P_p(0)}{P_{\text{sat}}^{ap}} - \sigma_{as}\right)N_tA_s(V)z\right] \tag{8.50}$$

The main point of this result is that, with overlapping spectra, considerably more pump power is needed to overcome the absorptive loss experienced by the signal.

A further consideration concerns noise arising from amplified spontaneous emission (ASE). As the name implies, gain will occur not only for the signal, but also for spontaneous emission that is radiated down the length of the fiber. As a rule, the signal-to-noise ratio increases with gain and as the separation between pump and signal wavelengths increases [23]. When resonant pumping is used, the best results are achieved by pumping at 1.45 μm, that is, near the short wavelength edge of the absorption spectrum.

The general effects of overlapping spectra and ASE were studied using a detailed rate equation analysis [23], the results of which are shown in Fig. 8.9. In the figure, the small signal gain over a length L_{amp} of fiber amplifier is plotted as a function of wavelength for several pump wavelengths. Since the material and structural parameters of length, emission crossection, and doping level contribute to the gain, it is convenient to define a normalized length parameter [23]:

$$L_n \equiv L_{amp}A_s(V)\sigma_{es}(\lambda_s)N_t \tag{8.51}$$

In the plots of Fig. 8.9, $L_n = 9$ at $\lambda_s = 1.53$ μm.

Alternate pumping wavelengths can be used that correspond to transitions from the ground state to any of the other excited states that are shown in Fig. 8.7. For example, light input at a wavelength of 800 nm will boost population

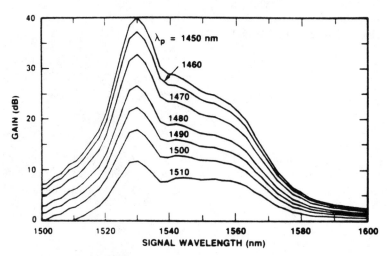

Figure 8.9. Calculated small signal gain for an erbium-doped amplifier using experimentally measured absorption and emission crossection data [23]. © 1991 IEEE.

from the ground state to the $^4I_{9/2}$ manifold, from which it rapidly decays to the metastable $^4I_{13/2}$ state to produce gain at 1.55 μm as before. The disadvantage in using the shorter pump wavelengths is that as population accumulates in $^4I_{13/2}$, further pump input can result in excited state absorption; in this process population is elevated from $^4I_{13/2}$ into the higher-energy states, thus depleting the population inversion and reducing the efficiency. The right-hand column in Fig. 8.7 indicates the wavelengths at which this will occur. In the case of pumping at 800 nm, it is seen that excited state absorption to the $^2H_{11/2}$ state (790 nm) competes with the gain-forming process and, for this reason, makes use of the 800 nm wavelength undesirable.

Among the possible choices, the only pump wavelength (aside from 1.48 μm) at which no excited state absorption from $^4I_{13/2}$ will occur is 980 nm. For the latter wavelength, the $^4I_{15/2} \rightarrow {}^4I_{11/2}$ transition is excited, followed by rapid decay to $^4I_{13/2}$. The peak absorption crossection for the transition is comparable to that at 1.48 μm. In contrast to the latter case, however, stimulated emission at 980 nm is completely absent ($\sigma_{ep} = 0$). Thus a larger fraction of the population excited by 980 nm pump light is able to thermalize within the $^4I_{13/2}$ state (and is thus available for gain at 1.55 μm) than is possible using 1.48 μm pump light at an equal power level. Photon conversion efficiencies from pump to signal of up to 80% have been achieved using this pump wavelength. In addition, pumping at 980 nm yields the largest signal-to-noise ratios obtainable, but these are in fact comparable to those achieved when pumping at 1.45 μm [21,23].

An amplifier scheme is shown in Fig. 8.10. The pump energy from the 1.48 μm or 980 nm wavelength diode laser is introduced into the amplifying fiber by way of a directional coupler. Critical to the operation of such a scheme is that pump energy must be applied continually to at least overcome the loss that the amplifier section presents with no pump. In the absence of pump energy, significant loss occurs due to the $^4I_{15/2} \rightarrow {}^4I_{13/2}$ transition. A typical loss spectrum for an erbium-doped fiber with no pump is shown in Fig. 8.11. The nonpumped erbium fiber in fact behaves as a saturable absorber at wavelengths in the vicinity of 1.5 μm, since as signal power increases, absorption (and consequent loss) decreases as the populations of the two states begin to equalize.

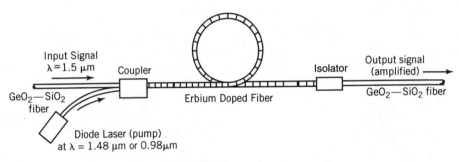

Figure 8.10. EDFA pumping arrangement.

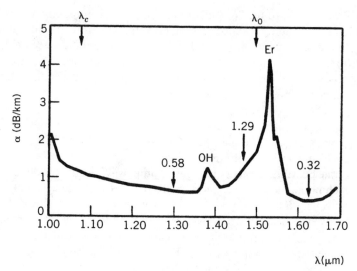

Figure 8.11.   Loss spectrum of a 6.3 km long EDFA with no pumping [24]. © 1991 IEEE.

The fabrication of erbium-doped fibers is accomplished using the same techniques discussed earlier. In the MCVD process, erbium ions are introduced into the preform deposition layers along with the other constituent materials that control refractive index. One of the major problems is that Er^{3+} ions tend to "cluster" in the glass matrix, leading to interactions between adjacent ions. This is a severe detriment to gain formation, since various nonradiative decay processes that depend on ion–ion interaction will occur [21]. The clustering problem has been eliminated completely through the addition of alumina (Al_2O_3) to the glass layers. This material increases the solubility of individual erbium ions and thus enables their incorporation into the glass host in a uniform distribution of high density. Germania GeO_2) is added as before to provide control of the waveguiding properties. An attractive feature of the fabrication process is that the erbium ion concentration can be adjusted from layer to layer to accommodate the shape of the mode profile. For example, the greatest concentration can be located at the center of the core, thus situating the greatest amplifier capacity where the mode intensity is highest.

PROBLEMS

8.1.   Using (8.12) and (8.13), derive an expression for the required pulse area to achieve an $N = 1$ soliton in terms of the operating wavelength, dispersion parameter D, nonlinear index n_2', mode effective area A_{eff}, and other known parameters as necessary.

8.2. Determine the required peak power to propagate a 10 ps width pulse as a soliton. Take $\lambda_m = 1.55\mu$m, at which the fiber dispersion is $D = 16$ ps/nm-km. Assume $A_{eff} = 5 \times 10^{-7}$ cm^2.

8.3. Estimate the pump power at the critical condition for stimulated Raman scattering for a 30 km single-mode fiber link for the following cases: (a) operation at 1.3 μm wavelength, at which the loss is 0.5 dB/km; and (b) operation at 1.55 μm at which the loss is 0.2 dB/km. Assume cw operation and a mode effective area of $A_{eff} = 5 \times 10^{-7}$ cm^2.

8.4. Transform-limited pulses of center wavelength $\lambda_2 = 1.55$ μm and 10 ps widths are transmitted over the fiber link of Problem 8.3. Using representative dispersion data from Chapter 5, determine the walkoff distance, L_w, and then estimate the critical power for stimulated Raman scattering.

8.5. For the fiber link of Problem 8.3, estimate the pump power at the critical condition for stimulated Brillouin scattering.

8.6. Using (8.45), (8.46), and (8.49), and assuming weak pump and signal powers, derive (8.50). Further assumptions are $N_1 \approx N_t$, $P_s \ll P_p$, and pump absorption is small, such that $P_p \approx P_p(0)$.

8.7. Determine an expression for the maximum ratio of N_3 to N_t obtainable in a material possessing absorption *and* emission crossections at the pump wavelength (assume the signal power is zero).

8.8. For zero signal power estimate the pump power required to equalize the populations of the $^4I_{13/2}$ and $^4I_{15/2}$ manifolds at $z = 0$ in an erbium-doped fiber where the pump wavelength is (a) $\lambda = 0.98\,\mu$m and (b) $\lambda = 1.48\,\mu$m. Use $\sigma_{ap}(0.98\,\mu$m$) = 3.1 \times 10^{-21}$ cm^2 (in Al/P silica) and obtain other crossections from Fig. 8.10 as needed. Also assume that all of the population excited to level $^4I_{11/2}$ relaxes to level $^4I_{13/2}$. Typical values for other parameters (some of which may be relevant) are $N_t = 10^{18}$ cm^{-3}, $A_c = 2 \times 10^{-7}$ cm^2, $A_p(V) \approx A_s(V) = 0.4$, and $\tau = 10$ ms.

REFERENCES

1. M. Stern, J. P. Heritage, R. N. Thurston, and S. Tu, "Self-Phase Modulation and Dispersion in High Data Rate Fiber-Optic Transmission Systems," *IEEE Journal of Lightwave Technology*, vol. 8, pp. 1009–1015, 1990.

2. R. H. Stolen and C. Lin, "Self-Phase Modulation in Silica Optical Fibers," *Physical Review A*, vol. 17, pp. 1448–1453, 1978.

3. G. P. Agrawal, *Nonlinear Fiber Optics*. Academic Press, Orlando, FL, 1989.

4. A. Hasegawa and F. Tappert, "Transmission of Stationary Nonlinear Optical Pulses in Dispersive Dielectric Fibers," *Applied Physics Letters*, vol. 23, pp. 171–172, 1973.

5. C. V. Shank, R. L. Fork, R. Yen, R. H. Stolen, and W. J. Tomlinson, "Compression of Femtosecond Optical Pulses," *Applied Physics Letters*, vol. 40, pp. 761–763, 1982.

6. R. L. Fork, O. E. Martinez, and J. P. Gordon, "Negative Dispersion Using Pairs of Prisms," *Optics Letters*, vol. 9, pp. 150–152, 1984.

7. R. L. Fork, C. H. Brito Cruz, P. C. Becker, and C. V. Shank, "Compression of Optical Pulses to Six Femtoseconds Using Cubic Phase Compensation," *Optics Letters*, vol. 12, pp. 483–485, 1987.

8. G. Herzberg, *Infra-red and Raman Spectra of Polyatomic Molecules*. Van Nostrand, New York, 1945, pp. 99–101.

9. F. L. Galeener and G. Lucovsky, "Longitudinal Optical Vibrations in Glasses: GeO_2 and SiO_2," *Physical Review Letters*, vol. 37, pp. 1474–1478, 1976.

10. R. H. Stolen and E. P. Ippen, "Raman Gain in Glass Optical Waveguides," *Applied Physics Letters*, vol. 22, pp. 276–278, 1973.

11. R. H. Stolen, "Nonlinearity in Fiber Transmission," *Proceedings of the IEEE*, vol. 68, pp. 1232–1236, 1980.

12. F. L. Galeener, J. C. Mikkelsen, Jr., R. H. Geils, and W. J. Mosby, "The Relative Raman Cross Sections of Vitreous SiO_2, GeO_2, B_2O_3, and P_2O_5," *Applied Physics Letters*, vol. 32, pp. 34–36, 1978.

13. R. G. Smith, "Optical Power Handling Capacity of Low Loss Optical Fibers as Determined by Stimulated Raman and Brillouin Scattering," *Applied Optics*, vol. 11, no. 11, pp. 2489–2494, 1972.

14. J. Auyeung and A. Yariv, "Spontaneous and Stimulated Raman Scattering in Long Low Loss Fibers," *IEEE Journal of Quantum Electronics*, vol. QE-14, no. 5, pp. 347–352, 1978.

15. R. H. Stolen and A. M Johnson, "The Effect of Pulse Walkoff on Stimulated Raman Scattering in Fibers," *IEEE Journal of Quantum Electronics*, vol. QE-22, pp. 2154–2160, 1986.

16. L. G. Cohen and C. Lin, "A Universal Fiber-Optic (UFO) Measurement System Based on a Near-IR Fiber Raman Laser," *IEEE Journal of Quantum Electronics*, vol. QE-14, no. 11, pp. 855–859, 1978.

17. A. R. Chraplyvy, "Optical Power Limits in Multi-channel Wavelength-Division Multiplexed Systems Due to Stimulated Raman Scattering," *Electronics Letters*, vol. 20, pp. 58–59, 1984.

18. R. H. Stolen, "Nonlinear Properties of Optical Fibers," in *Optical Fiber Telecommunications*, S. E. Miller and A. G. Chynoweth, eds. Academic Press, New York, 1979.

19. A. R. Chraplyvy, "Limitations on Lightwave Communications Imposed by Optical Fiber Nonlinearities," *IEEE Journal of Lightwave Technology*, vol. 8, pp. 1548–1557, 1990.

20. A. Corney, *Atomic and Laser Spectroscopy.* Clarendon Press, Oxford, 1977, pp. 83–86.

21. W. J. Miniscalco, "Erbium-Doped Glasses for Fiber Amplifiers at 1500 nm," *IEEE Journal of Lightwave Technology*, vol. 9, pp. 234–250, 1991.

22. D. E. McCumber, "Theory of Phonon-Terminated Optical Masers," *Physical Review*, vol. 134, pp. A299–A306, 1964.

23. C. R. Giles and E. Desurvire, "Propagation of Signal and Noise in Concatenated Erbium-Doped Fiber Optical Amplifiers," *IEEE Journal of Lightwave Technology*, vol. 9, pp. 147–154, 1991.

24. J. R. Simpson, et al., "Performance of a Distributed Erbium-Doped Dispersion-Shifted Fiber Amplifier," *IEEE Journal of Lightwave Technology*, vol. 9, pp. 228–233, 1991.

Appendix A

Properties of Bessel Functions

In solving the wave equation for the fields in a step index fiber, the differential equation for the radially dependent portion of the product solution was found to be a form of the Bessel equation:

$$\frac{d^2R(r)}{dr^2} + \frac{1}{r}\frac{dR(r)}{dr} + \left(\beta_t^2 - \frac{n^2}{r^2}\right)R(r) = 0 \tag{A.1}$$

where β_t and n are constants.

The Bessel equation has two classes of solutions, depending on whether β_t is real or imaginary. In the former case, the general solution consists of ordinary Bessel functions of the first and second kinds of order n and of argument $\beta_t r$:

$$R(r) = AJ_n(\beta_t r) + BN_n(\beta_t r) \tag{A.2}$$

where

$$J_n(\beta_t r) = \sum_{p=0}^{\infty}\frac{(-1)^p(\beta_t r/2)^{n+2p}}{p!\,\Gamma(p+n+1)} \tag{A.3}$$

and

$$N_n(\beta_t r) = \frac{J_n(\beta_t r)\cos(n\pi) - J_{-n}(\beta_t r)}{\sin(n\pi)} \tag{A.4}$$

The constant n need not be an integer, although in the case of the optical fiber it is restricted to integer values, in which case $\Gamma(p+n+1) = (p+n)!$. For integer n, J_n can also be expressed in integral form:

$$J_n(\beta_t r) = \frac{1}{\pi}\int_0^{\pi}\cos(\beta_t r\phi - n\phi)\,d\phi \tag{A.5}$$

Plots of J_n and N_n for integer n are shown in Fig. 3.3. Zeros of J_n are given in Table A.1.

Functions having positive and negative order are related through

$$J_{-n}(\beta_t r) = (-1)^n J_n(\beta_t r) \tag{A.6}$$

$$N_{-n}(\beta_t r) = (-1)^n N_n(\beta_t r) \tag{A.7}$$

Recall that a complex exponential can be represented in terms of sine and cosine functions by means of the Euler equation:

$$\exp(\pm jkx) = \cos(kx) \pm j\sin(kx) \tag{A.8}$$

In analogy, the *Hankel functions* of the first and second kind are defined in terms of the ordinary Bessel functions through

$$H_n^{(1)}(\beta_t r) = J_n(\beta_t r) + jN_n(\beta_t r) \tag{A.9}$$

$$H_n^{(2)}(\beta_t r) = J_n(\beta_t r) - jN_n(\beta_t r) \tag{A.10}$$

Large argument forms ($\beta_t r \to \infty$) of the above functions for integer n are:

$$J_n(\beta_t r) \to \sqrt{\frac{2}{\pi \beta_t r}} \cos\left(\beta_t r - \frac{\pi}{4} - \frac{n\pi}{2}\right) \tag{A.11}$$

$$N_n(\beta_t r) \to \sqrt{\frac{2}{\pi \beta_t r}} \sin\left(\beta_t r - \frac{\pi}{4} - \frac{n\pi}{2}\right) \tag{A.12}$$

Table A.1 Zeros of $J_n(z)$

J_0	J_1	J_2	J_3	J_4	J_5	J_6	J_7
2.405	3.832	5.136	6.380	7.588	8.771	9.936	11.086
5.520	7.016	8.417	9.761	11.065	12.339	13.589	14.821
8.654	10.173	11.620	13.015	14.373	15.700	17.004	18.288
11.792	13.324	14.796	16.223	17.616	18.980	20.321	21.642
14.931	16.471	17.960	19.409	20.827	22.218	23.586	24.935

$$H_n^{(1)}(\beta_t r) \rightarrow \sqrt{\frac{2}{\pi \beta_t r}} \exp\left[j\left(\beta_t r - \frac{\pi}{4} - \frac{n\pi}{2} \right)\right] \qquad (A.13)$$

$$H_n^{(2)}(\beta_t r) \rightarrow \sqrt{\frac{2}{\pi \beta_t r}} \exp\left[-j\left(\beta_t r - \frac{\pi}{4} - \frac{n\pi}{2} \right)\right] \qquad (A.14)$$

Small argument forms ($\beta_t r \ll 1$) for integer n are

$$J_n(\beta_t r) \approx \frac{1}{n!} \left(\frac{\beta_t r}{2} \right)^n \qquad (A.15)$$

$$K_n(\beta_t r) \approx (n - 1)!\, 2^{(n-1)}(\beta_t r)^{(-n)} \qquad (A.16)$$

A Bessel or Hankel function of a given kind can be expressed in terms of others of the same kind but of different order through the recurrence relation:

$$\frac{2n}{\beta_t r} F_n(\beta_t r) = F_{n+1}(\beta_t r) + F_{n-1}(\beta_t r) \qquad (A.17)$$

where F_n represents J_n, N_n, or H_n.

Derivatives of Bessel and Hankel functions can be expressed in terms of nondifferentiated functions through

$$F'_0(\beta_t r) = -F_1(\beta_t r) \qquad (A.18)$$

$$F'_1(\beta_t r) = F_0(\beta_t r) - \frac{1}{\beta_t r} F_1(\beta_t r) \qquad (A.19)$$

$$\beta_t r F'_n(\beta_t r) = n F_n(\beta_t r) - \beta_t r F_{n+1}(\beta_t r) \qquad (A.20)$$

$$\beta_t r F'_n(\beta_t r) = -n F_n(\beta_t r) + \beta_t r F_{n-1}(\beta_t r) \qquad (A.21)$$

where the derivatives are taken with respect to the *total argument*.

The second class of solutions to (A.1) results when β_t is an imaginary number. These, known as modified Bessel functions, comprise the general solution, written in the form

$$R(r) = C K_n(|\beta_t| r) + D I_n(|\beta_t| r) \qquad (A.22)$$

The new functions are defined in terms of the ordinary Bessel functions through

$$I_{\pm n}(|\beta_t|r) = j^{\pm n} J_{\pm n}(j\beta_t r) \tag{A.23}$$

$$K_n(|\beta_t|r) = \frac{\pi}{2} j^{(n+1)} H_n^{(1)}(j\beta_t r) \tag{A.24}$$

The first two integer orders of K_n and I_n are sketched in Fig. 3.4. Large argument forms of the modified Bessel functions are

$$j^{-n} J_n(j\beta_t r) \approx I_n(|\beta_t|r) \rightarrow \sqrt{\frac{1}{2\pi|\beta_t|r}} \, \exp(|\beta_t|r) \tag{A.25}$$

$$j^{n+1} H_n^{(1)}(j\beta_t r) \approx \frac{2}{\pi} K_n(|\beta_t|r) \rightarrow \sqrt{\frac{2}{\pi|\beta_t|r}} \, \exp(-|\beta_t|r) \tag{A.26}$$

Recurrence formulas for the functions are

$$\frac{2n}{|\beta_t|r} I_n(|\beta_t|r) = I_{n-1}(|\beta_t|r) - I_{n+1}(|\beta_t|r) \tag{A.27}$$

$$\frac{2n}{|\beta_t|r} K_n(|\beta_t|r) = K_{n+1}(|\beta_t|r) - K_{n-1}(|\beta_t|r) \tag{A.28}$$

Derivatives (again taken with respect to the total argument) are

$$|\beta_t|r I'_n(|\beta_t|r) = n I_n(|\beta_t|r) + |\beta_t|r I_{n+1}(|\beta_t|r) \tag{A.29}$$

$$|\beta_t|r I'_n(|\beta_t|r) = -n I_n(|\beta_t|r) + |\beta_t|r I_{n-1}(|\beta_t|r) \tag{A.30}$$

$$|\beta_t|r K'_n(|\beta_t|r) = n K_n(|\beta_t|r) - |\beta_t|r K_{n+1}(|\beta_t|r) \tag{A.31}$$

$$|\beta_t|r K'_n(|\beta_t|r) = -n K_n(|\beta_t|r) - |\beta_t|r K_{n-1}(|\beta_t|r) \tag{A.32}$$

REFERENCES

1. E. Jahnke and F. Emde, *Tables of Functions*. Dover, New York, 1945)
2. M. Abramowitz and I. Stegun, *Handbook of Mathematical Functions*. Dover, New York, 1965.
3. S. Ramo, J. R. Whinnery, and T. Van Duzer, *Fields and Waves in Communication Electronics*, 3rd ed. John Wiley & Sons, New York, 1994.

Appendix B

Notation

The following is a list of terms that appear frequently throughout the text. Along with the description of each parameter, the equation number where the term is first described (where appropriate) is given in parentheses.

a	Fiber core radius
A	Sellmeier coefficient (5.23)
A_c	Fiber core cross-sectional area
A_s	Surface area of light source
A_{eff}	Effective mode crossectional area (8.4)
$A(V)$	Fraction of LP_{01} mode power that exists within the core of a step index fiber (5.35), (5.45)
b	Normalized propagation constant (3.78)
B	Photometric brightness (4.30), Rayleigh scattering coefficient (4.9), bit rate
B_{eff}	Effective bandwidth in stimulated Raman scattering (8.14)
$\mathbf{B}$	Phasor form of magnetic induction (tesla)
$\mathcal{B}$	Real instantaneous form of magnetic induction
c	Velocity of light in vacuum
c.c.	Complex conjugate
d	Slab waveguide film thickness
$D(\lambda)$	Fiber dispersion parameter (5.25)
D_m	Material dispersion (5.42)
D_w	Waveguide dispersion (5.43)
D_p	Profile dispersion (5.44)
$\mathbf{D}$	Phasor form of electric displacement (C/m^2)
$\mathcal{D}$	Real instantaneous form of electric displacement
e	Electron charge
$\mathbf{E}$	Phasor form of electric field strength (V/m) (1.12)
$\mathcal{E}$	Real instantaneous form of electric field strength (1.10)
f	Frequency (Hz), focal length
g_r	Raman gain coefficient (7.84)
g_b	Brillouin gain coefficient (7.114)
$\mathbf{H}$	Phasor form of magnetic field strength (A/m)
$\mathcal{H}$	Real instantaneous form of magnetic field strength
$I_n(w)$	Modified Bessel function having order n and argument w

$J_n(u)$	Ordinary Bessel function of the first kind, having order n and argument u
k	Propagation constant of a plane wave in a bulk material having refractive index n: $k = nk_0 = n\omega/c$
k_0	Free-space plane wave propagation constant (ω/c)
k_{x1}	Transverse (x-directed) propagation constant in a slab waveguide film (2.3)
k_{ijk}	Nonlinear "spring constant" (7.9)
$K_n(w)$	Modified Bessel function having order n and argument w
l	Azimuthal mode number for LP modes (3.45)
L	Fiber length
L_D	Dispersion length parameter (8.7)
L_{NL}	Nonlinear length parameter (8.8)
L_{eff}	Effective length parameter (8.15)
L_w	Walkoff length (8.23)
m	Mode number of slabguide (2.47), radial mode number of fiber
$m(\beta)$	Number of modes in a graded index fiber that have azimuthal mode number, l, and that have propagation constant values between β and β_{max} (6.22)
m	Mass of a particle
M	Lens magnification (4.36)
n	Refractive index
n_{eff}	Effective index for a guided mode (β/k_0) (3.78)
n_1	Refractive index in core of step index fiber or dielectric slab waveguide
n_2	Refractive index in cladding of step index fiber or dielectric slab waveguide
n_2'	Nonlinear refractive index parameter (7.44)
$n(r)$	Radial refractive index function (6.2)
N	Group index of a material (5.22)
N_f	Net group index (including material, waveguide, and profile dispersion) of a fiber (5.50)
N_i	Population density in level i in an optical amplifier
$N_n(u)$	Ordinary Bessel function of the second kind, having order n and argument u
$N.A.$	Numerical aperture (3.2)
P	Net power flow (flux of $\mathbf{S}$)
$\mathcal{P}$	Polarization of the medium (real instantaneous) (1.6)
$\mathcal{P}_L$	Linear medium polarization (7.5)
$\mathcal{P}_{NL}$	Nonlinear medium polarization (7.5)
q	Azimuthal mode number for TE, TM, and hybrid modes (3.10), (3.11)
q	Number of longitudinal modes within Raman bandwidth (8.14)
q_i	Real instantaneous displacement in ith direction
r	Radial variable in cylindrical coodinates

r_b	Outer bend radius of slab waveguide
r_c	Critical bend radius of slab waveguide
r_m	Bend radius of slab or fiber axis
r_0	Mode field radius (3.100), (6.65); minimum spot radius of gaussian beam
r_1, r_2	Ray turning point radii in a graded index fiber
$r_g(z)$	z-dependent gaussian beam radius
$R(r)$	Radially dependent part of the optical fiber field product solution (3.8), (3.13)
$\mathbf{S}$	Poynting vector (W/m²) (1.22)
$s(r)$	Scalar function that describes mode phase distribution in a graded index fiber (gradient gives ray trajectory) (6.12)
t	Time
t_g	Pulse group delay over a unit distance
t_L	Local time variable within pulse envelope
T	Gaussian pulse halfwidth at $1/e$ (5.1)
u	Normalized radial propagation constant in fiber core (3.14)
v	Phase velocity (ω/β)
v_g	Group velocity ($d\omega/d\beta$)
v_p	Acoustic wave velocity in Brillouin scattering
V	Normalized frequency parameter (3.46)
V_c	Value of V at cutoff for a given fiber mode
w	Normalized radial mode decay coefficient in a step index fiber (3.15)
x, y	Transverse cartesian variables
y	Profile dispersion parameter (5.38)
z	Mode propagation direction
z_0	Rayleigh range of a focused gaussian beam (3.104)
α	Graded index profile parameter (6.2), loss coefficient
α_b	Macrobending exponential power loss coefficient (4.16)
α_c	Power loss from coupling (4.29)
α_d	Microbending exponential power loss coefficient (4.25)
α_{IR}	dB/km power loss from infrared absorption (4.7)
α_p	dB/km power loss coefficient (Problem 1.8), (4.1); acoustic wave attenuation coefficient
$\alpha_p(z, t)$	Time-varying polarizability (7.64)
α_s	Rayleigh scattering coefficient in (dB/km-μm⁴) (4.9)
α_{UV}	dB/km power loss from ultraviolet absorption
β	Mode propagation constant, z component of plane wave propagation constant
β_t	Net transverse component of ray propagation constant in a fiber
γ	Electrostrictive coefficient
γ_2	Transverse exponential decay coefficient for slab waveguide fields (2.9)
Γ_{TE}	Reflection coefficient for TE waves at a dielectric interface (1.44)

Γ_{TM}	Reflection coefficient for TM waves at a dielectric interface (1.46)
δ	Normalized detuning from resonance (1.57)
Δ	Normalized index difference (3.1)
Δk	Phase mismatch per unit distance (7.36)
$\Delta\tau$	Pulse spread in halfwidth (5.6)
$\Delta\omega_s$	Halfwidth of source power spectrum (5.14)
$\Delta\omega_{\mathrm{eff}}$	Effective signal bandwidth (5.17)
$\Delta\omega_b$	Brillouin linewidth
$\Delta\omega_r$	Raman linewidth
$\Delta\lambda_{\mathrm{eff}}$	Effective bandwidth (5.27)
ϵ	Dielectric permittivity (1.5)
ϵ_0	Free-space permittivity (1.5)
ζ	Damping coefficient (linewidth parameter) (1.52)
η	Fiber coupling efficiency (4.28), instrinsic impedance of a bulk material
η_0	Intrinsic impedance of free space
θ	Altitude variable in spherical coordinates
θ_B	Brewster's angle of total transmission (1.49)
θ_c	Critical angle of total reflection (1.51)
$\theta_{N.A.}$	Numerical aperture angle ($\sin^{-1}(N.A.)$)
λ	Free-space wavelength of light
λ_m	Mean wavelength in a power spectrum
λ_0	Zero-dispersion wavelength
λ_c	Cutoff wavelength (3.87)
μ	Magnetic permeability
μ_0	Free-space permeability
$\nu(\beta)$	Number of modes in a graded index fiber between β and β_{max} (6.26)
ν_T	Total number of modes in a graded index fiber (6.31)
ρ	Mass density
ρ_v	Volume charge density
σ	Conductivity, rms pulse width ($T/\sqrt{2}$)
σ_τ	rms pulse spread (5.61)
σ_{ap}	Absorption crossection at pump wavelength
σ_{as}	Absorption crossection at signal wavelength
σ_{ep}	Emission crossection at pump wavelength
σ_{es}	Emission crossection at signal wavelength
τ	Characteristic decay time for spontaneous emission (8.29)
τ_g	Group delay of a pulse propagating over distance z (5.19)
ϕ	Azimuth variable in cylindrical and spherical coordinates
ϕ_{TE}	One-half the phase shift by total reflection of a TE wave (2.14)
ϕ_{TM}	One-half the phase shift by total reflection of a TM wave (2.15)
$\Phi(\phi)$	ϕ-dependent part of the optical fiber field product solution (3.8), (3.12)
χ_e	Electric susceptibility function (1.6), (1.55)

$\chi^{(1)}$	Linear susceptibility (scalar or tensor)
$\chi^{(2)}$	Second order susceptibility (scalar or tensor)
$\chi^{(3)}$	Third order susceptibility (scalar or tensor)
$\chi^{(2)}_{ijk}$	Second order susceptibility tensor element (e.g., 7.10)
$\chi^{(3)}_{ijkl}$	Third order susceptibility tensor element (e.g., 7.12)
$\chi_{(1)}$	Off-diagonal third order susceptibility tensor element for an isotropic medium (along with $\chi_{(2)}$ and $\chi_{(3)}$) (7.19) to (7.21)
χ_r	Raman susceptibility function (7.79)
ψ_b	Brillouin coupling parameter (Problem 7.12)
ψ_r	Raman coupling parameter (7.83), (8.18)
ω	Radian frequency ($2\pi f$)
ω_0	Radian frequency of pulse carrier (5.1)

Index

6-13-95